THE DVB-H HANDBOOK
THE FUNCTIONING AND PLANNING OF MOBILE TV

THE DVB-H HANDBOOK

THE FUNCTIONING AND PLANNING OF MOBILE TV

Jyrki T. J. Penttinen
Nokia Siemens Networks, Spain

Petri Jolma
Nokia Siemens Networks, Finland

Erkki Aaltonen and Jani Väre
Both of Nokia, Finland

WILEY

A John Wiley and Sons, Ltd, Publication

This edition first published 2009
© 2009 John Wiley & Sons Ltd

Registered office
John Wiley & Sons Ltd, The Atrium, Southern Gate, Chichester, West Sussex, PO19 8SQ,
United Kingdom

For details of our global editorial offices, for customer services and for information about how to apply for
permission to reuse the copyright material in this book please see our website at www.wiley.com.

Library of Congress Cataloging-in-Publication Data

The DVB-H handbook : the functioning and planning of mobile TV / Jyrki Penttinen ... [et al.].
    p. cm.
  Includes bibliographical references and index.
  ISBN 978-0-470-74829-9 (cloth)
  1. DVB-H (Standard) 2. Mobile television–Planning. I. Penttinen, Jyrki, 1967-
  TK6678.5.D84D84 2010
  621.388'5–dc22

                                                                    2009033733

A catalogue record for this book is available from the British Library.

ISBN: 978-0-470-74829-9 (H/B)

Set in 11/13 pt, Times Roman by Thomson Digital, Noida, India
Printed in Great Britain by CPI Antony Rowe, Chippenham, Wiltshire

Contents

About the Authors

Mr. Jyrki T.J. Penttinen has worked in the area of telecommunications since 1994, for Telecom Finland and it's successors until 2004, and after that for Nokia and Nokia Siemens Networks. He has performed various international tasks e.g. as a System Expert and Senior Network Architect in Finland, R&D Manager in Spain and Technical Manager in Mexico and USA. He currently holds a Senior Solutions Architect position in Madrid, Spain. His main activities have been related to mobile and DVB-H network design and optimization.

Mr. Penttinen obtained the M.Sc. (E.E.) and Licentiate of Technology (E.E.) degrees from Helsinki University of Technology (TKK) in 1994 and 1999, respectively. He has organized various telecom courses and lectures. In addition, he has published various technical books and articles since 1996.

Mr. Jani Väre obtained M.Sc. degree in Information Technology from Tampere University of Technology (TUT) in 2002 and he is currently finalizing his PhD thesis. He has worked for Nokia since 2001, participating closely in the activities related to Mobile TV and DVB-H. Throughout his Nokia career he has contributed to the generation of new technologies within the field of

Telecommunications and especially related to Mobile TV. The development of IPDC over DVB-H and DVB-T2 standards has been his latest major projects. Currently, he is working closely in the development of the Next Generation Handheld standard, which is the latest evolution of Mobile TV within DVB. He holds over 40 patents and patent applications. He is also named in the Marquis Who's Who in Science and Engineering and in the Marquis Who's Who in the World.

As Senior Manager, New Business Development at Nokia Sales and Industry Collaboration, **Erkki Aaltonen** has a broad range of responsibilities that cover sales and marketing as well as strategy and business development in the Mobile TV business area. During his long career with Nokia, he has been pivotal in identifying strategic new business ideas and stimulating long-term growth opportunities. In addition to roles in Nokia, Erkki has been involved in various start-ups and ventures.

Erkki holds a master's degree in economics from the University of Tampere, School of Economics and Business Administration. In his spare time, he is a keen football player and also enjoys travelling, Italian cuisine and spending time with his family.

Petri Jolma received the degree of M.S. of Electrical Engineering from the Technical University of Helsinki in 1986, and has been employed by Nokia Networks/Nokia Siemens Networks since then. He has worked at several positions in microwave design, cellular network RF design and research, including DVB-H technology.

Preface

Digital Video Broadcasting, Handheld (DVB-H) is an efficient solution for the delivery of broadcast type information in the mobile environment. The system has been developed based on the terrestrial version of digital television by adding essential functionalities for the challenging radio environment of hand-held terminals. As the system has been standardized via European Telecommunications Standards Institute (ETSI) in 2004, the solutions and equipment are ready to be installed to the live environment.

This book presents the technical and commercial principles, functionality and planning aspects of DVB-H. The book is structured in modules. The first part includes the overall revision of the system and its position in the related market field (Chapters 1, 2 and 12). The second part explains the business models and marketing-related issues of DVB-H (Chapter 3). The third part goes into the details of the

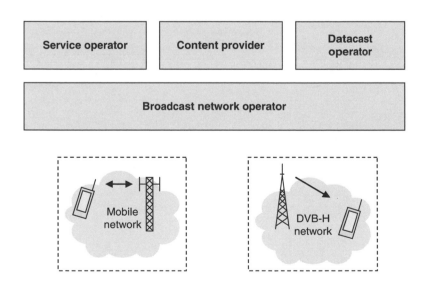

Figure 1 The complete view of the DVB-H environment

DVB-H network functionality, including the head-end system with related core elements, interfaces and signalling (Chapters 4–8). The fourth part presents a detailed DVB-H radio network planning and optimization aspect (Chapters 9–11), which can be utilized in the deployment and quality revisions of the system.

This book is aimed for broadcast and mobile network operators, network and terminal manufacturers, service providers, frequency regulators and technical students of universities and colleges as a study material for related telecommunication courses.

It should be noted that this book represents the views and opinions of the authors, and does not necessarily represent the views of their employers.

Acknowledgements

The author team would like to acknowledge the professional support of our colleagues in contributing to this book. Besides our core group of authors (Jyrki T.J. Penttinen, Jani Väre, Erkki Aaltonen and Petri Jolma), the other contributors are Eric Kroon (Chapters 1, 2 and 5), Juha Viinikainen (Chapter 8) and Kyösti Koivisto (Chapter 7) to whom we want to express our gratitude. In addition, there were various colleagues giving valuable feedback during the work. The team would like to give special thanks to Jukka Henriksson and Pekka Talmola for the most helpful support.

The processing of the information presented in this book has taken logically longer time for each author than the actual writing period lasted. The authors are grateful for all the professional support of our colleagues from various companies years before this book was written, including the standardization and pilot phases of DVB-H. The team would like to express gratitude for all the interchange of ideas – regardless of the company or organizational boundaries – to Jouni Oksanen, Pekka Pesari, Anssi Korkeakoski, Stefan Schneiders, Otto Afflerbach, Ari Savolainen, Clay Simmons, Johan Backa, Samur Worasilpchai, Steve Crisler, Lynn Leffler, Jason Banaag, Dario Ambrosini, Paul Hartman, and all the others that we have been working with in different corners of the world.

It was a pleasure to work with the team at John Wiley & Sons participating in the production of this book. The professional support was of great help in keeping the project in schedule. Special thanks go to Mark Hammond, Sarah Hinton, Sophia Travis, Sarah Tilley and Katharine Unwin for the practical assistance, as well as Brett Wells for the shaping of the expressions of the non-native speakers of the contributors.

It would have been much more difficult to carry out the work without the patience and understanding of the close ones of the team. The authors would like to give special thanks all their family members.

The team welcomes any comments that might improve the next editions of this book. The feedback is welcome to the e-mail address jyrki.penttinen@nsn.com.

Abbreviations

16-QAM	16-state Quadrature Amplitude Modulation
2G	Second Generation (of mobile networks)
3G	Third Generation (of mobile networks)
64-QAM	64-state Quadrature Amplitude Modulation
AAA	Authentication, Authorisation and Accounting
AAC	Advanced Audio Coding
HE-AAC	High Efficiency Advanced Audio Coding
ADT	Application Data Table
AES	Audio Engineering Society
ALC	Asynchronous Layered Coding
ARP	Address Resolution Protocol
ARPU	Average Return Per User
ASC	Address Scrambling Control
ASI	Asynchronous Serial Interface
AU	Access Unit
BAM	Broadcast Account Manager
BC	Broadcaster/Content Providers
BCAST	Mobile Broadcast Services Enabler Suite
BER	Bit Error Rate (before Viterbi)
BNO	Broadcast Network Operator
BSM	Broadcast Service Manager
CA	Conditional Access
CAPEX	Capital Expenses
CBMS	Convergence of Broadcast and Mobile Services
CBR	Constant Bit Rate
CENELEC	European Committee for Electrotechnical Standardization
CIF	Common Intermediate Format (176×220 pixels)
CMLA	Content Management License Administration
COST	European Co-operation in the Field of Scientific and Technical Research

CP	(1) Content Provider, (2) Cyclic Prefix
CPE	Common Phase Error
CR	Code Rate
CRC	Cyclic Redundancy Check
CRM	Customer Relationship Management
CTS	Composition Time Stamp
dBi	Decibels compared to isotropic antenna
DCO	Datacast Operator
DRM	Digital Rights Management
DTS	Decoding Time Stamp
DVB	Digital Video Broadcasting
DVB-C	Digital Video Broadcasting, Cable
DVB-CBMS	Convergence of Broadcasting and Mobile Service
DVB-H	Digital Video Broadcasting, Handheld
DVB-NGH	Digital Video Broadcasting, Next Generation Handheld (previously referred as DVB-H2)
DVB-S	Digital Video Broadcasting, Satellite
DVB-T	Digital Video Broadcasting, Terrestrial
EBU	European Broadcasting Union
ECMG	Entitlement Control Message Generator
EIRP	Effective Isotropically Radiated Power (reference being omni antenna)
EMC	Electromagnetic Compatibility
EPG	Electronic Program Guide (see ESG)
ERP	Effective Radiated Power (reference being half-wavelength dipole antenna)
ESG	Electronic Service Guide
ETSI	European Telecommunications Standardization Institute
FEC	Forward Error Correction
FER	Frame Error Rate (before MPE-FEC)
FFT	Fast Fourier Transform
FLUTE	File Delivery over Unidirectional Transport
GI	Guard Interval
GPRS	General Packet Radio Service
GPS	Global Positioning System
GSM 850	GSM in 850 MHz band
GSM 900	GSM in 900 MHz band
GSM	Global System for Mobile Communications
HE	(1) Head-End, (2) High Efficiency
HP	High Priority
HW	Hardware
I/Q	In-phase and Quadrature Plane

ICI	Inter-Carrier Interference
ICMP	Internet Control Message Protocol
IETF	Internet Engineering Task Force
IGMP	Internet Group Management Protocol
IMSI	International Mobile Subscriber Identity
IMT	International Mobile Telecommunications
INT	IP/MAC Notification Table
IPDC	IP Datacast
IPE	IP Encapsulator
IPsec	IP Security Architecture
IPv4	Internet Protocol version 4
IPv6	Internet Protocol version 6
ISDB-T	Integrated Services Digital Broadcasting, Terrestrial
ISDN	Integrated Services Digital Network
ISI	Inter-Symbol Interference
ISMA	Internet Streaming Media Alliance
ISMACryp	ISMA Encryption and Authentication
ISO	International Standardization Organization
ITU	International Telecommunication Union
IV	Initialization Vector
LAN	Local Area Network
LP	Low Priority
LTE	Long Term Evolution
MBMS	Mobile Broadcast Multicast Service
MBS	Mobile Broadcast Solution
MCC	Mobile Country Code
MCS	Modulation and Coding Scheme
MFER	Frame Error Rate (after MPE-FEC)
MFN	Multi-Frequency Network
MIP	Megaframe Initialization Packet
MMS	Multimedia Message Service
MNC	Mobile Network Code
MNO	Mobile Network Operator
MPE	Multi-Protocol Encapsulation
MPE-FEC	Multi-Protocol Encapsulation – Forward Error Correction
MPEG	Moving Picture Experts Group
MSISDN	Mobile Subscriber ISDN Number
NIT	Network Information Table
NO	Network Operator
OMA	Open Mobile Alliance
OPEX	Operating Expenses
OSI	Open Standards Interface (ISO definition)

PAL	Phase Alternating Line
PAT	Program Association Table
PER	Packet Error Ratio
PID	Packet Identifier
PLR	Packet Loss Ratio
PMT	Program Map Table
PPS	Packets per Second
PSC	Payload Scrambling Control
PSI	Program Specific Information
RAP	Random Access Point
QCIF	Quarter Common Intermediate Format (176×144 pixels)
QEF	Quasi Error Free (quality criteria)
QoS	Quality of Service
QPSK	Quadraphase-Shift Keying
QVGA	Quarter Video Graphics Array (320×240 pixels)
RF	Radio Frequency
RO	Rights Object
RSSI	Received Signal Strength Indicator
RTCP	RTP Control Protocol
RTP	Real Time Protocol
RSTP	Rapid Spanning Tree Protocol
RTT	Finnish Radio and Television Investigation Society
SDI	Serial Digital Interface
SDP	Service Discovery Protocol
SFN	Single Frequency Network
SI	Service Information
SIM	Subscriber Identity Module
SMS	Short Message Service
SNR	Signal-to-Noise Ratio
SO	Service Operator
SRTP	Secure Real-time Transport Protocol
STKM	Short Term Key Message
SW	Software
TDD	Time Division Duplexing
TDT	Time and Date Table
TPS	Transmission Parameter Signaling
TS	Transport Stream
TU	Typical Urban (for propagation models)
UDP	User Datagram Protocol
UHF	Ultra-High Frequency
UMTS	Universal Mobile Telecommunications System
USIM	Universal Services Identity Module

VBER	Bit Error Rate (after Viterbi)
VBR	Variable Bit Rate
VGA	Video Graphics Array
VHF	Very High Frequency
WLAN	Wireless Local Area Network
XML	Extensible Markup Language

List of Contributors

Kyösti Koivisto, Nokia Siemens Networks, Finland

Eric Kroon, Nokia Siemens Networks, Finland

Juha Viinikainen, Nokia Siemens Networks, Finland

List of Figures

1

General

1.1 Setting up the Scene

The development of wireless communication technologies has changed our living style in global level. After the international success of mobile telephony standards, the forerunner being GSM, the location and time independent voice connection has become a default method in daily telecommunications. The Short Message Service (SMS) came into the picture in the mid-1990s. As for today, whether the messaging is based on the simple but robust short message service or highly advanced multimedia messaging, it plays a key role in value-added service handling. Along with evolving data services, the need for more complex applications can be seen, including the mobile usage of broadcast technologies.

The real commercial outbreak of the television became true in the 1950s. Since then, the television has been a permanent piece of furniture in every household. During the long life of television, there have been basically only two major evolution steps: introduction of colours and digital transmission. The logical next step is the mobility. Although portable analogue TV sets have been available long time ago, it is challenging to maintain their reception quality in the mobile environment.

The Japanese ISDB-T was the first digital television standard which offered possibility of viewing television with the handsets such as mobile phones with satisfactory end user experience. The European Digital Video Broadcasting (DVB) standard was developed soon after this. Today, the DVB standards consist of several different standards covering terrestrial, satellite and cable systems. The

The DVB-H Handbook   Jyrki T.J. Penttinen, Erkki Aaltonen, Jani Väre and Petri Jolma
© 2009 John Wiley & Sons, Ltd

first complete DVB-H specification was published in 2004 which generated various trials and pilots. One of the first DVB-H trials was set up in Finland during 2004. Approximately at the same time, the DVB-CBMS sub-group was founded in DVB to define the IPDC over DVB-H standard, which would specify a complete end-to-end solution for the delivery of multimedia services over DVB-H. The DVB-IPDC system layer specifications cover such aspects as Electronic Service Guides, Content Download Protocols and Service Purchase and Protection.

It is not an overstatement to say that, despite that the ISDB-T was the first standard in introducing Television service, i.e. MobileTV to the handheld devices, the DVB-H and especially OMA-BCAST over DVB-H has been the pioneer in making the global outbreak for MobileTV. Today the DVB-H has spread all the way from Europe to Africa, Asia, Australia and Latin America, leaving out only the North America and some other markets like Japan and South Korea. Italy was the first country to launch commercial DVB-H services in 2007. Several commercial launches have followed since then. DVB-H is also supported by European Union as the dominant digital television system for handhelds.

Whilst the broadcast technologies have emerged to the stage where consumers can view broadcast transmissions through their handhelds, such as mobile phone, also the cellular systems have reached the point where almost the same content and services can be offered through e.g. 3G and also soon over further advanced systems such as Long Term Evolution (LTE). One could now ask whether this makes sense. Could we live only with either one?

1.2   Benefits of DVB-H

Mobile TV is a method that provides the users with high channel capacity and interactivity 'on the road'. It brings totally new aspects on the personal information handling, whether it is about leisure time with entertaining TV program clips or complicated business solutions. This interactive mobile multimedia is one of the key ideas of the next steps of multimedia era.

The parallel usage of two-way communication and broadcasted content brings new challenges to the building, operating and optimizing the networks that deliver the services. Mobile TV is a service that provides the users with many TV programs and interactivity on the move. It brings totally new aspects on the personal information consumption, whether it is about leisure time with entertaining TV programs, short video clips or informative type of service. Interactive mobile multimedia is one of the key ideas of the next steps of multimedia era, and broadcast component provided with DVB-H (Digital Video Broadcasting – Handheld) is a useful addition to the conventional cellular radio networks serving the users with point-to-point connections.

Clear advantage of the mobile broadcast system compared to the cellular system-based services is that it provides the service at large areas without capacity limitations at receiving end. Any user within the coverage area can receive the service simultaneously, however numerous they are. Total amount of received bits can be several decades higher than with point-to-point cellular technologies. Nevertheless, the mobile TV service might not be the 'killer application' as a stand-alone solution, but combination of broadcast and point-to-point cellular data and voice networks brings the best synergy, increasing the total usage more than the simple sum of individual network usage would be.

The DVB-H adds features to the DVB-T which improve the performance of handheld terminals and mobile reception: the first the time slicing feature (a form of time division multiplexing) which enables to reduce considerably the power consumption and enables seamless handovers. The second feature, the MPE-FEC (Multi Protocol Encapsulated–Forward Error Correction) improves the threshold C/I ratio which gives adequate reception quality. Also it increases the tolerance to Doppler shift and impulse interference. The third feature, the 4K FFT gives additional flexibility to network planning especially when terminals are travelling on high speed vehicles.

With the DVB-H radio carrier, which has the same bandwidth and is practically the same as DVB-T carrier, it is possible to transmit several video and/or audio streams due to the low bit rate of MPEG-4 video coding. A typical DVB-H service carries more TV programs than DVB-T service, even though the DVB-H carrier has lower total bit rate than DVB-T. The small screen size of mobile or portable receivers makes this possible. The bit rate of mobile TV programs ranges typically from 100 to 400 kb/s, only approximately 1/10 what is typical at terrestrial DVB-T, around 4–5 Mb/s. A typical DVB-H carrier has the total bit rate 1/2 to 1/4 of a DVB-T carrier due to differently set channel coding parameters; still the DVB-H carrier may contain more programs than a DVB-T carrier due to the low bit rate of programs.

The DVB-H solves the non-optimal aspects of DVB-T that are related in the moving environment. First one is the power consumption, which can be reduced considerably in DVB-H due to the time slicing solution. The performance in cellular environment in general requires a strong carrier per noise rate in, which is in sufficiently decent level in DVB-H even if the Doppler in mobile channel and impulse interferences are present. DVB-H also provides network design flexibility for the mobile with a single antenna reception in medium- to large-scale single frequency network areas.

The full-scale DVB-H is actually a combination of broadcast and mobile operations, with respective mobile broadcast key representatives that are Network Operator (NO), Datacast Operator (DCO), Content Provider (CP) and Service Operator (SO). It is interesting to note that there are several ways to create the business model and related setup of the roles.

1.3 Standardization

DVB-H belongs to the DVB family, the other variants being DVB-T (terrestrial), DVB-C (cable) and DVB-S (satellite). DVB-H is based on the DVB-T and is backwards compatible with it. Handheld mobile TV service motivated a creation of a new DVB standard. During the DVB-T standardization mobility was already taken into account so that the basic mobility capability was already there, but some additions made the standard more beneficial. It can be said that the DVB-H is the DVB-T standard tailored to meet the requirements and needs of handheld mobile devices. In the DVB-H band, which would be useful for a single DVB-T channel at the time, it is possible to transmit several DVB-H sub-channels containing video and/or audio due to the lower capacity need per DVB-T channel and smaller screen size of the terminal.

The DVB-H specification was planned in technical and commercial modules, i.e. sub-groups, of the DVB project, which is a co-operative effort of ETSI (European Telecommunications Standardization Institute), EBU (European Broadcasting Union) and CENELEC (European Committee for Electrotechnical Standardization). The system specification was included into ETSI specifications, and the first version was published in 2004, resulting various trials and pilots. One of the first trials was initiated in Finland during 2004, and Italy was the country where the first launch of commercial DVB-H service took place in 2007. The system is also supported by European Union. The system has not stayed specific to Europe, but it has been evaluated as a strong candidate in other continents. As the telecommunications systems tend to evolve for higher capacities, performance enhancement is planned also for the DVB-H. The next generation is called DVB-NGH, previously referred to as DVB-H2. The main goal is to provide more capacity, and more flexibility for the service bit rate.

It can be said that the DVB-H is the DVB-T standard tailored to meet the requirements and needs of handheld devices. Similar to DVB-T and all other standards of the 'DVB family of standards', DVB-H supports transport stream. The main differences in DVB-H vs. DVB-T are in the power consumption, error protection and signalling. Another fundamental difference between DVB-T and DVB-H is that the terminal specifics are taken into account, light weight, portable and battery powered devices.

The DVB-H specification, which is limited to the radio transmission, is not sufficiently complete to enable global interoperability for different networks and devices. In 'DVB project', which created the DVB standards, the DVB-CBMS (Convergence of Broadcasting and Mobile Service) sub-group was founded to define IPDC (IP datacast) over DVB-H standard, which specified a complete end-to-end solution for the delivery of multimedia services over DVB-H. The DVB-IPDC system layer specifications cover such aspects as Electronic Service Guides, Content Download Protocols and Service Purchase and Protection. The DVB-H

Figure 1.1   The logotype of the DVB-H system*

thus forms the radio layers, i.e. physical and media access layers, for larger entity of Mobile Broadcast system.

While the DVB-H standard defines the OSI layer 1 and 2 protocols, the IPDC over DVB-H standard defines the protocols up to the OSI layers 3–7. In addition, the IPDC over DVB-H standard defines also some extensions and rules for the DVB-H specific Program Specific Information (PSI)/Service Information (SI) signalling. During the development process of the IPDC over DVB-H standard, an alternative solution for the OSI layers 3–7 over DVB-H was defined by the Open Mobile Alliance (OMA). This OMA BCAST solution adopts great part of the IPDC over DVB standard, including PSI/SI as such. The other way round, the IPDC over DVB-H standard has also adopted some parts from the OMA BCAST. Today the roles of the OMA BCAST and IPDC over DVB-H standards have become clearer. The OMA-BCAST defines the upper layer solution and the IPDC over DVB-H standard defines adaptation for the DVB-H bearer. This 'OMA-BCAST over DVB-H' is today supported by the most of the mobile operators and device vendors which provide products for the DVB-H-based mobile TV systems. Figure 1.1 shows the official logo of DVB-H.

As the telecommunication systems tend to meet the requirements of higher capacity, also DVB-H includes its evolution path. The next generation of the DVB-H system is called DVB-NGH (which was previously referred to as DVB-H2). The main idea of it is to provide more capacity optimizing the coverage area. According to the convergence of the cellular and fixed communication systems, DVB-H is expected to be part of a complete set of multimedia systems.

1.4   Contents of the Book

This book presents a complete set of topics related to the background of DVB-H, its business models, technical functionality, network planning and optimization. It can be used as a technical reference material in the planning and operational phases of the DVB-H network.

The book has been divided into three modules. The first module presents the overall DVB-H concept with the respective business environment, explaining the

---

*The DVB and DVB stylized marks, either used in association with the DVB sub-brand logos, or not, are registered trademarks of the DVB Project.

high level solutions and operating models for different parties of the whole DVB-H business chain. The related topics are presented in Chapters 1–3.

The second part goes into the details of the service functionality, including the signalling and setting up of the service. This module is presented in Chapters 4–7.

The last module presents the detailed principles of the core and radio network functionality with the dimensioning aspects of the network, including also the in-depth optimization items. This module is presented in Chapters 8–11.

Chapter 2 presents an overview of the DVB-H. The chapter gives a short introduction to the standardization of the system, as well as information about the other mobile broadcast systems. Chapter 3 describes models and roles for different parties in the mobile TV business, including the mobile operator, virtual network operator, wholeseller and reseller, broadcast network operator and variations where roles are combined.

Chapter 4 explains the architecture of the DVB-H network for both core and radio network sides. In this book, the term core network means the network part from the program encoder elements to the IP encapsulator, including all the related elements that take care of the transport within the internal delivery paths. The radio network of DVB-H takes care of encapsulating the IP packets to the MPEG-2 transport stream, the transport of the MPEG-2 transport stream to the radio transmitter sites and DVB-H modulation and radio transmission. This chapter also describes the interfaces and protocols used in communication between the elements. Chapter 5 goes into the physical aspects and characteristics of the DVB-H equipment in the core and radio network. It also gives information about the measurement equipment that can be used as a part of the radio and core network planning and optimization. Chapter 6 describes in-depth the functionality of the DVB-H.

Chapter 7 contains information about the signalling of DVB-H. More specifically, it gives an overview to service discovery signalling, explaining how the service can be found in a certain frequency band. The chapter continues with the presentation of the functionality when the terminal is used, including the handovers for the smooth service continuity then travelling. It also describes shortly the issues related to the parallel interaction channel. The DVB-H specifications apply only to downlink direction of the radio network. For the possible interactions, basically any means of transferring messages from the customer's side can be applied, including 2G, 3G and other two-way networks. The most logical solution is to use GPRS (General Packet Radio Service) functionality of the mobile networks as it is typically integrated into DVB-H handsets. The additional benefit of using 2G and 3G is that they already contain all the needed methods for the authentication, authorization and accounting (AAA) of the customers, so e.g. the opening of the protected contents can be done via packet data connection in a controlled way by using the IMSI (International Mobile Subscriber Identity) of 2G and 3G in DVB-H interactions.

In addition, the chapter shows digital rights-related issues of DVB-H. There is a possibility of delivering both free-to-air and closed contents. The latter requires special measures in order to keep the content rights and related commercial issues in order.

Chapter 8 discusses the core network planning with the in-depth revision of the network architecture and dimensioning of the head-end and transport network, i.e. the core chain from the encoders up to the IP encapsulator. Chapter 9 presents the radio-related parameters and functionality. Chapter 10 presents the detailed radio network dimensioning aspects. Chapter 11 presents the means to optimize the DVB-H network with relevant examples.

Finally, Chapter 12 discusses about the evolution of the DVB-H with probable solutions in the near future.

2

DVB-H Overview

The DVB-H is a broadcast system, which was designed in a backward compatible way, based on the DVB-T. The DVB-H solves the non-optimal points of DVB-T, which are related in the moving environment. The core features of the DVB-H are time slicing and Multiprotocol Encapsulation Forward Error Correction (MPE-FEC). The time slicing decreases the power consumption, while the MPE-FEC improves robustness of the data transmission. In addition to these two features, also additional 4K FFT mode, DVB-H specific signalling extensions in PSI/SI and enhanced TPS signalling are new features of DVB-H as compared to DVB-T.

2.1 Time Slicing

Time slicing is based on the principle where the receiver power can be set to off-state periodically when receiving the services. The periodical reception of data is enabled by the 'bursty' transmission of data, where there are off-periods with no data transmission and periods when the data is transmitted. These off-periods are signalled within real-time parameters, which are carried within the MPE section headers. In addition, PSI/SI signalling also contains time slicing specific signalling. The time slicing principle is depicted in Figure 2.1, where the signalling within the MPE section headers indicates the time, i.e. delta-t, to the start of the next burst. The time slicing signalling is explained with more detail in Chapter 6.

Time slicing means that the audio/video stream or other content is sent in high bit rate burst. The DVB-H receiver stores the data in an in-built buffer and starts streaming the contents whilst the receiver is switched off. The receiver is basically

The DVB-H Handbook   Jyrki T.J. Penttinen, Erkki Aaltonen, Jani Väre and Petri Jolma
© 2009 John Wiley & Sons, Ltd

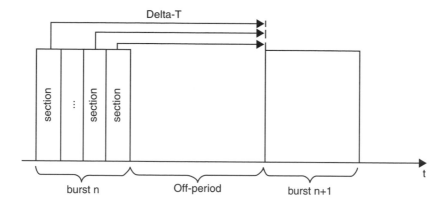

Figure 2.1   The time slicing principle

switched on only during the reception of the burst that the terminal is currently following. This provides about 90% power saving as far as the receiver is concern.

In addition to the power saving, the time slicing also enables seamless handover with single tuner and front end. The principle for the seamless handover within DVB-H is based on the utilization of off-periods for inspection of available neighbouring cells. The seamless handover is synonym to 'soft handover'. It was first introduced in [Var04].

2.2   MPE-FEC

MPE-FEC was developed to improve robustness of data transmission in the reception environments where there is high packet loss ratio (PLR) on the MPE section level. High PLR occurs typically on mobile channels when the receiver is in a high speed and/or the C/N is too low.

MPE-FEC is the enhancement which is not mandatory to be used by the receiver, even if it is transmitted by the network. Hence, the receiver may ignore MPE-FEC and still be able to receive the application data carried within the MPE sections.

MPE-FEC can be allocated separately for each elementary stream. Hence, the network operator may not use MPE-FEC on all elementary streams e.g. if the robustness level is considered to be in satisfactory level without the use of MPE-FEC. The use of MPE-FEC is always a trade-off between consumption of network capacity and robustness. Another trade-off could be seen between the end-to-end delay and the robustness level of the time-critical services, such as live sport events where slight errors in the transmission could be more acceptable than delay in the transmission. MPE-FEC is dealt with greater detail in Chapter 6.

2.3   DVB-H Specific Signalling

The DVB-H specific signalling can be split into two levels: upper layer signalling and DVB-H bearer specific signalling. The upper layer signalling, which consists of OMA-BCAST specific signalling, relates to the ESG and content delivery protocols such as FLUTE. The DVB-H bearer specific signalling consists of PSI/SI signalling and TPS signalling which are carried in OSI layers 2 and 1.

The DVB-H specific signalling consists of PSI/SI signalling and TPS signalling. The PSI/SI signalling and the TPS signalling within DVB-H are based on the legacy signalling used within the DVB-T. However, new fields have been added to the TPS signalling for the use of DVB-H. Also the set of PSI/SI tables has been constraint only to those which are relevant for DVB-H. The PSI/SI tables used within the DVB-H are Program Association Table (PAT), Program Map Table (PMT), Network Information Table (NIT), IP/MAC Notification Table (INT) and Time and Date Table (TDT). The DVB-H specific signalling and its use within the DVB-H service discovery are dealt exhaustively in Chapter 7.

2.4   The Broadcast and Cellular Systems in DVB-H

It can be said that the broadcast systems and cellular systems are from two different worlds. The broadcast network's commitment to the service delivery ends when it has transmitted the service on the air. In addition, the broadcast network does not have any information about the terminals which are consuming the services that it is providing. The cellular network, in turn, is active throughout the duration of the service delivery. It is also aware of the terminals which are consuming the service offering and maintains communication with these. Finally, in the mobile environment e.g. during the handovers, the cellular network controls the handover procedure. In the broadcast system, only the signalling is provided by the network and the terminal is fully responsible for handling the handover.

Even though the core functionality within the DVB-H is based on pure broadcasting, the complete end-to-end DVB-H system, such as OMA-BCAST over DVB-H also utilizes the interaction network that can be provided by the cellular systems, such as GSM or 3G. It can be said that DVB-H is a combination of the broadcast and cellular systems. For some services and use cases, DVB-H is seen as pure broadcast. An example of the latter is the free-to-air services, where the content is free, no encryption is used and hence no interaction between the network and end-user terminals is needed. The role of cellular systems within DVB-H, or more preferably OMA-BCAST over DVB-H, is in the interactivity needed between the network and the end user mainly due to service and content protection related communication and signalling. Also, some service scenarios, such as the ones including chatting or voting, require interactivity.

2.5 Market Needs

DVB-H provides a solution for the growing demand of having more and more demanding multimedia 'on the road'. The video contents can be delivered via the mobile communication networks via point-to-point circuit-switched data transmission methods, packet data transfer, multimedia messages, or e.g. mobile broadcast multicast service (MBMS). Each one of these methods has their limitations, though, compared to the dedicated broadcast systems like DVB-H.

Circuit-switched data occupies the channel permanently until it is released, which is the least efficient in the transmission of the vastly varying capacity of audio and video. Packet-switched data, as GPRS, is optimal for the bursty IP packets generated by the audio and video contents as the system utilizes mainly the 'leftover' capacity that is almost always available in correctly dimensioned voice service networks. Nevertheless, there is also a capacity limit in this solution especially when the amount of required channels gets higher. This can be noted as a clear service level degradation especially in busy hour of the mobile network.

MBMS is a broadcast service over the already existing mobile networks. It can utilize the separately defined capacity block, and the rest of the traffic continues as before for the voice and data services. The total capacity in the existing 2G and 3G mobile networks is limited though to some megabits per second, and when sufficiently large part of the existing capacity is dedicated to MBMS, the voice and data service capacity or service level gets lower. For this reason, MBMS might not be the optimal solution for the broadcast traffic, although the service can reutilize directly the existing network infrastructure. The evolution of the mobile networks offers higher data rates in the future, which makes MBMS a more logical solution for the capacity, but there are other aspects that DVB-H copes better than in the MBMS environment as seen below.

DVB-H is a separate network although the mobile network sites can be physically reutilized at least partially. It is highly probable that DVB-H needs anyway own dedicated sites and also higher masts in order to offer sufficiently large service areas. Also the transmission is separate in DVB-H. Depending on the available site transmission capacity, the same physical infrastructure can also be utilized for DVB-H.

DVB-H is optimized for the portable and pedestrian type of environments due to the enhancements in the radio reception, i.e. due to the error-coding capabilities. This means that the terminal speed is not a limiting factor in the normal user environment, whereas the Doppler shift might affect e.g. on the DVB-T reception by increasing the error rate.

One important aspect of the DVB-H receiver is the battery-saving functionality. It is based on the time slicing, which basically means that the terminal receives a high data rate burst in reduced time period and streams the selected contents by using the buffering. This makes it possible to switch off the DVB-H receiver for major part of time. Compared to the continuous streaming with the receiver all the time, the time saving and thus battery duration enhancement can be achieved up to 90–95%. It should be noted though that in addition to the receiver power consumption, there are also terminal-related functionality that requires energy all the time, like the video streamer application and the related screen with colour dot display. Nevertheless, the useful time for DVB-H audio/video reception can be normally around 4 h, i.e. clearly longer than that in circuit or packet-switched data reception.

The screen size of the DVB-H receiver is normally small e.g. 320×240 pixels (QVGA). This allows the optimization of the capacity as the individual video stream does not need to occupy more than 200–400 kb/s in order to provide a good user experience. The balancing of the capacity and data rates vs. frame rate is one of the various network dimensioning tasks that are described in more detail in Chapters 9 and 10.

2.6 Applications

DVB-H provides a means to deliver various different services ranging from regular television and radio services to the file delivery and even services supporting interactivity.

2.6.1 Television Service

The television service, i.e. video and audio, has been the core application within the DVB-H systems from the beginning.

2.6.2 Radio Service

In addition to the television type of reception, DVB-H can be utilized for the radio reception as well. In the practice radio service is from the technical point of view similar to the television service; only the video part is missing. The DVB-H coding method for the audio is AAC, and with some 64 kb/s it is possible to provide a high-quality stereophonic transmission. This would be suitable for listening to e.g. classical music. Depending on the contents, the data rate can be considerably lower, which is the case e.g. in the voice news type of services.

2.6.3 File Delivery

Even though the DVB-H is a unidirectional broadcast system; it is also possible to deliver data files via the DVB-H channels. This can be done e.g. by defining a carousel type of service for a part of the channels. The service repeats some specific contents, i.e. one or several different files that can be received via the DVB-H terminal, and used either via the same DVB-H equipment or via e.g. laptop computer. The file contents are not technically limited as the transmission is purely data, and the files can thus be web pages, text documents, multimedia contents, etc.

2.6.4 Special Solutions

As DVB-H is a system offering services and coverage areas that are somewhere between the mobile networks and television/radio networks, it is also suitable for many special services. One example could be an alarm delivery about local, national, or international events. The local contents can be delivered in the parts that belong to the same DVB-H core cell, covering e.g. a single city area. The emergency service could thus be very useful for informing e.g. about natural disasters, severe weather warnings, and accidents that can possibly affect the safety within the area. The strength of the DVB-H is the possibility to inform about the events in real time simply switching the emergency messages replacing the ongoing contents, with the description of the event as well as the safety instructions.

3

Mobile TV Business Eco-System

In this chapter, we will discuss the mobile TV business. We start by explaining the network of dependencies in mobile TV business, that is, what kinds of factors influence each others in the business, and how this may influence the consumer offering. After that we go through the generic mobile TV value chain, and explain what kinds of roles are needed for the business. We will also cover the most common mobile TV business and revenue models: mobile operator driven; vertical and virtual network operator business models, broadcast network operator driven; wholesaler–reseller business model, and broadcaster driven; NewCo business model. At the end of this chapter we will highlight the key factors that influence the DVB-H network investments. Regardless of what role different players want to take in the value chain, everyone should have an understanding of the network investment and the related operating costs.

3.1 The Network of Dependencies in Mobile TV Business

In mobile TV or any other consumer business, everything starts from the consumer needs. The following section discusses the network of dependencies in mobile TV business, in other words, what is needed for successful mobile TV business and how different factors influence each other. Figure 3.1 illustrates this.

Without consumer demand there is no business. Consumers have different reasons and motivations for wanting to watch mobile TV. As in any other consumer business, consumers may go thru behavioral, intuitive, emotional or rational levels when choosing whether to watch mobile TV or not.

The DVB-H Handbook   Jyrki T.J. Penttinen, Erkki Aaltonen, Jani Väre and Petri Jolma
© 2009 John Wiley & Sons, Ltd

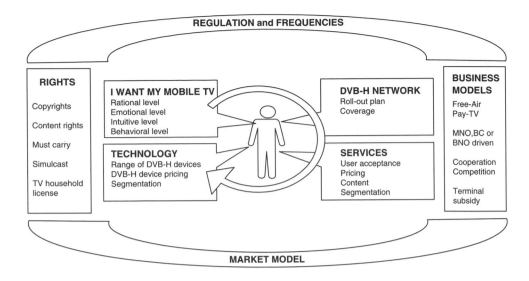

Figure 3.1 Network of dependencies in mobile TV business

After that consumers need to live in a network coverage area; otherwise they cannot consume the services even if they want to. The number of available frequencies in different areas determines where the operators can build the network coverage. Without frequencies there cannot be any network. Regulation and related network license terms may determine where the network should be build. Secondly this is linked to business models, and business cases, as the networks should be build only in areas where they make economic and business sense.

Consumer acceptance is of course linked to what kinds of services are offered, and at what price. Service providers decide which mobile TV channels and related services they would like to offer to their consumers. These services and the pricing must match the right consumer segments. This means right services for the right segments and at the right price. For example, if the consumer would like to watch music TV, but the service provider only has news channels, this would most likely influence the consumer interest. Or if the service provider offers music TV, but it costs X Euros per month and the consumer is expecting free service, again it is most likely that we see a change in the consumer interest level. Number of mobile TV channels offered may also influence interest level, as consumers prefer variety of choices. Especially in Pay TV services it is often easier to justify monthly X Euros if you get unlimited access to, e.g., 20 mobile TV channels, even if the user is actually interested in just a couple of those channels (we call this "SKY effect").[1]

---

[1] In the UK, satellite pay TV operator SKY is offering hundreds of TV channels. For many SKY's exclusive rights for Premier League Football is the biggest reason for subscribing the services, but for many consumers it is easier to justify the monthly fee as they get hundreds of channels on top of the football channel.

So business models and the service provider's preferences may influence the consumer interest.

Similarly content rights, copyrights, "must carry" rules and content-related regulation in a certain country will influence the service portfolio. These may have a direct impact on the related costs of offering the services to the consumers. For example, content rights for some very popular mobile TV channels may be quite expensive, but at the same time some less popular channels may actually pay to be part of the service portfolio. Operators also need to consider that the content rights for free-to-air channels are different than that for Pay TV channels. In some countries, regulator forces service providers to include public service channels as part of the service portfolio. Also, in many countries, there is also a debate about whether there is need to pay additional content rights for simulcast DVB-H mobile TV channels, if the rights are already paid for DVB-T. Discussion is going around also about the need for the separate TV license to watch mobile TV. Most of the regulators interpret the legislation so that current TV license paid by the consumer for his/her normal TV also covers mobile TV. All of these have impact on business models and service portfolios and therefore consumer interest.

As well as the segmentation for the services having to be right, the segmentation must also be right for the available DVB-H devices. Certain consumer segments demand certain types of phones at certain price levels. It might influence the consumer interest level if the available devices are not for the right segment, even if the consumer is offered the best mobile TV services at an optimal price. For example, if the operator is only offering business user phones, the services should match this segment and vice versa. Pricing and availability of the devices can be influenced by operator subsidies, but this is of course dependent on the business model and operator role. In the case of the business model, that bypassed the mobile operators, the service providers have limited ways to influence the available devices unless they control the device distribution in that market.

Regulation and available frequencies also influence the potential business models in the country. For example, if there are limited number of frequencies available only for one national network, then the regulator may force the companies to cooperate to avoid harmful monopoly situation. A very common case is where one broadcast network operator (BNO) is awarded with the network license, but they have to sell the capacity on equal terms to several service providers. It also means that one service provider is not allowed to buy 100% of the network capacity. The regulator may also set a number of conditions in which companies are allowed to operate in certain business areas. For example, in some countries special media license is needed to aggregate mobile TV service portfolio. Regulation may also force the separation of content and networks, so the network operator is not allowed to aggregate content and act as service provider.

Business models and what kind of role different companies can and are willing or allowed to take in the value chain have probably the biggest impact on mobile

TV business. As mentioned above, there are several areas that influence the business (e.g., regulation, frequencies, content related), so the players are not always free to choose what kind of role they want to take in the mobile TV business or what kind of services they can offer and at what price. This may influence the interest levels of the potential key stakeholders: mobile operators (MNOs), broadcasters/content providers (BC), and broadcast network operators (BNOs). The key question is whether the different companies will co-operate or compete against each other when building the business. For example, mobile operators are natural service providers in any mobile business, as they already have a direct consumer relationship. They also subsidize mobile devices in many markets. If mobile operators are bypassed in the mobile TV business, this would require, e.g., setting up new forms of customer care, and most likely would mean that there would not be any device subsidies. This might influence the range of available DVB-H devices and the interest level of consumers. But then again business model driven by the broadcaster might lead to free-to-air services which again can influence consumer interest levels, if the only other option would be pay TV services. Also in some markets the regulator does not allow any other company than those with media or broadcasting license to act as service providers in that country, leading to a case where mobile operators may struggle to find a suitable role in the value chain.

All of the above, regulation, frequencies, content, business models, consumer interest, and network, will determine the market model for the country. Market model summarizes all the elements of mobile TV business, based on whether there is equilibrium found in a market or not. It also indicates how successful the market will be. Perfect equilibrium means that all players can co-operate in a win–win situation, and the optimum number of services can be offered at the optimum price to the consumers. But as it is very rare that all players would have exactly the same view on the market and especially to have the same negotiation power, the equilibrium will be shaped accordingly. This may lead that some of the players (e.g., one of the mobile operators in the country or some TV channels) decide not to enter the business, as it does not offer enough incentive, or they lack negotiation power or funds to enter the business. For other players this can open the market model where they can dominate even more and make even higher profits.

3.2 Mobile TV Business Roles

The mobile TV business eco-system includes various different roles, but the implementation varies from market to market. Mobile TV offers real business opportunities for the whole business eco-system including mobile operators, service providers, broadcasters, content providers, device and network element vendors and broadcast network operators amongst others.

All business models may have several variations depending on the market and in all markets each of the key stakeholders (mobile operator, broadcaster, and broadcast network operator) may take different roles in value chain. In theory, one key stakeholder could take all key roles (building the DVB-H network, aggregating mobile TV service portfolio and offering the service to consumers), but in practice stakeholders current core competences and risk taking abilities lead to cooperation models between the players. It is also possible, and in some cases even likely, that different business models, or their variations, coexist in the market.

Each of the roles may have several companies acting in them, for example, hundreds of content providers, tens of TV channels, several mobile operators (typically two to four per country), several mobile TV channel aggregators, and in some cases also more than one broadcast network operator depending on frequency of availability and if the business cases allow several networks. Figure 3.2 illustrates the main roles in mobile TV business value chain.

A mobile broadcast network operator builds and operates the DVB-H network. It typically owns the frequency or network license, but can also in some cases act on behalf of third party who owns the license. The most common business model for the broadcast network operator is based on a fixed fee for the raw network capacity, and it is not involved in content aggregation, or in direct consumer interface. In this model its revenue depends on how large network coverage and capacity it builds. Very often broadcast network operator has a monopoly in a country, which means that the regulator often sets the limits for the capacity pricing. Broadcast network operator may have contracts with several different players in the value chain depending on the business model. It can sell the capacity directly to mobile TV channels and other content providers, or the mobile TV channel aggregators, or to cellular network operators or some combination of these.

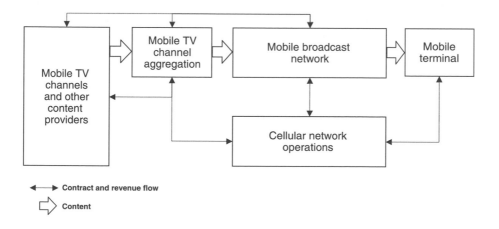

Figure 3.2   Mobile TV business opportunities

Mobile TV channel aggregator is a player who creates the service portfolio for the consumers. This is a very critical role, as the right service portfolio is essential for the success of the business. Mobile TV channel aggregator negotiates channel deals with the mobile TV channels (broadcasters) and other content providers. The channel aggregator also negotiates deals with service resellers (mobile operators) if the aggregator is not a part of the mobile operator's role. Very often the market model assumes that mobile TV channel aggregator(s) will also pay for the broadcast network capacity. So often, the player who takes this role will also drive the business and carry most of the risk, but of course most of the rewards as well. In some cases a special media license is required for this role.

Mobile TV channels (often called broadcasters) and other content providers are essential for the successful business. Without content there is no mobile TV business. The key role for the TV channels (broadcasters) is to aggregate individual programs from the content providers to create mobile TV channels. Mobile TV channels can of course simulcast channels to normal TV. In this case broadcasters just need to clear the content rights for mobile distribution before they can sell the channel to mobile TV aggregators.

Mobile operators can have several roles. As they already have an existing customer relationship, they are the natural CRM providers for the mobile TV business as well. Also marketing, customer acquisition, retention, device distribution, selling the service and billing and interactive services are natural for the mobile operators. Often mobile operators are seen as the most natural service providers for the mobile TV services, as mobile TV is business as usual for the operators, one service among other mobile services. In the case of broadcaster-driven, free-to-air business model, it is also possible that the mobile operator could be bypassed and excluded from the value chain. Whether that kind of business model will succeed is another question.

The device manufacturer's role is to provide attractive devices to the customers. The key is to provide a wide range of devices for different consumer segments and price bands. Devices are sold thru mobile operator distribution channels or thru independent distributors and retailers.

3.3   Vertical Business Model, Mobile Operator-Driven Model

In a vertical business model, the mobile operator builds and operates its own DVB-H network and aggregates own mobile TV service portfolio and offers that to the consumers. It may also own some content, but most likely not all. In a vertical business model, the mobile operator controls the entire value chain and creates the mobile TV offering to its own existing customers. It is close replicate of the mobile operators' existing business model in cellular environment. Figure 3.3 illustrates a vertical business model.

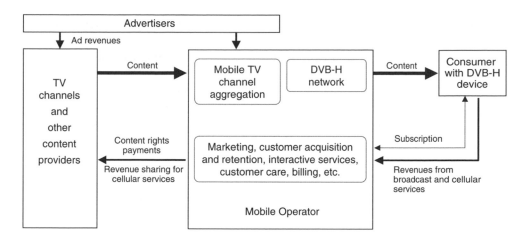

Figure 3.3   Vertical business model

The mobile operator owns the DVB-H network license and builds the coverage according to its own needs (as long as it is fulfilling the network license terms). It also negotiates content deals with TV channels (broadcasters) and other content providers. Operators create a service portfolio and promote it to consumers. They can offer different kinds of service bundles at different prices to specific customer segments. Also it is very common that mobile operators bundle mobile TV services with the existing cellular offerings. An operator can offer, e.g., 15 mobile TV channels and 500 SMS and 500 min of voice per month for X Euros.

In addition to creating service portfolio and operating the DVB-H network, the key activities for the operator are marketing, customer acquisition and retention, customer care and billing. Also in many cases mobile operators control device distribution, and can significantly influence the business by device subsidies.

Operators also provide the interaction channel, as cellular networks are used to deliver DRM keys for the encrypted Mobile TV services. Also one key part of the service offering is interactive services related to mobile TV, like voting or additional web information. Mobile operators create interactive services, together with third party developers, as they will trigger the usage of existing cellular networks, which means more revenues.

Revenue models may vary significantly between different vertical operators. Most common is that the operator offers part of the mobile TV channels free of charge to the consumers. These are either promotional TV channels, meaning that those make consumers want to try the service, or are popular channels funded by adverting or by interactive services. Free channels are also commonly used for customer acquisition; in this case the operator uses mobile TV to gain new mobile customers, not just mobile TV users. Part of the mobile TV channels is usually premium pay TV channels, meaning that consumers need to pay, e.g., a monthly fee to access those services.

Best real life example of a vertical business model is Hutchison 3G Italy. They have build their own DVB-H network, aggregated an attractive service portfolio (including exclusive rights for the Italian Series A Soccer) and have some own made for mobile content production. They are subsidizing DVB-H devices and bundling the mobile TV and cellular services to the consumers. At the end of 2008, Hutchison 3G Italy had around 1 million mobile TV customers.

3.4 Virtual Network Operator Model, Mobile Operator Operator-Driven Model

In a virtual network operator model, the mobile operator coordinates the activities by purchasing content rights for mobile TV channels from broadcasters and capacity from the broadcast network operator. The service is then offered to consumers in a similar way to that in the vertical business model. The difference from the mobile operator's perspective is that in the virtual network operator model, the operator does not own the DVB-H network or the network license. Another difference from the vertical model is that in virtual network operator model, it is likely that there are several mobile operators (or service providers) buying the capacity from one network. From the mobile operator's perspective, this model is close to MVNO (mobile virtual network operator) models familiar to mobile operators. Figure 3.4 illustrates the model.

Similar to that in the vertical model, the mobile operator also takes care of marketing, customer acquisition and retention, CRM, billing, interactivity, and in many cases device distribution.

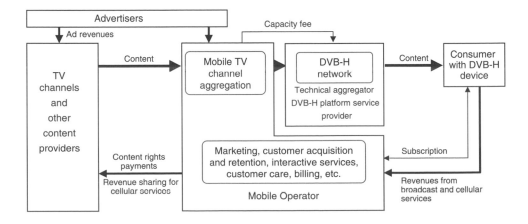

Figure 3.4   Virtual network operator model

Revenue models are very similar to the vertical business model. Also in the virtual network operator model, operators offer free channels based on advertising and interactivity and pay TV channels for subscription. Different flavors might come from the case where several mobile operators offer the same services. That might force the operator to a revenue model that it would not otherwise choose.

A real-life example of a vertical model is Austrian DVB-H market. In Austria, Media Broadcast owns the network license and sells the capacity to the service providers (mobile operators in Austria). The mobile operators Hutchison 3G Austria, Orange, and Mobilkom (A1) have negotiated content deals with 15 TV channels and 5 radio channels and are selling the service to their customers. In Austria all mobile operators are offering the same broadcast mobile TV services, but differentiate their offering with cellular service bundles.

In the virtual network operator model, the services do not have to be the same for all mobile operators, but in many cases, like in Austria, it helps co-marketing and saves DVB-H network capacity. Also the regulator might encourage service providers to cooperate. The same service portfolio requires strong cooperation between mobile operators, as in some cases it can be agreed which channels should be included in the service portfolio as mobile operators may have very different user segments. This is one of the biggest reasons why some operators either want to drive the content negotiation on behalf of all operators in the country or try to control the business by the vertical business model.

3.5 Wholesaler–Reseller Model, Broadcast Network Operator-Driven Model

The wholesaler–reseller model includes two dependent parts: a wholesaler and a service reseller. Wholesaler, typically a broadcast network operator, builds and operates the DVB-H network. In addition to this it also aggregates mobile TV service portfolio, which means they do the mobile TV channel line-up. The wholesaler offers this mobile TV service portfolio to the service resellers. The wholesaler is dependent on these service resellers (mobile operators), who sell the mobile TV service to consumers. So the wholesaler's success depends on how many service resellers and mobile operators it can sign in to sell the service. Figure 3.5 illustrates this.

In theory, wholesalers could bypass the mobile operators and either offer services free to air to consumers or offer pay TV services using conditional access systems. But this might significantly reduce the potential, as lack of mobile operators' support would risk the business case. As a wholesaler does not have a direct consumer relationship, it is essential that mobile operators are motivated to promote the services to consumers.

The wholesale model concentrates around an active broadcast network operator (wholesaler), who with its existing business relationships with media industry is

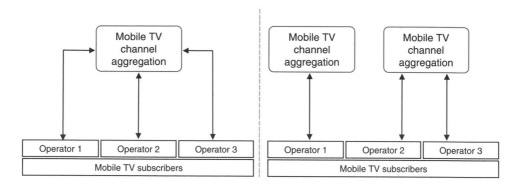

Figure 3.5 Wholesaler success depends on number of resellers

able to acquire the most relevant content from broadcasters and content owners and create a mobile TV service portfolio interesting enough for the mobile operator (service reseller) and consumers. Figure 3.6 illustrates the wholesaler–reseller model.

After buying the content the wholesaler sells the portfolio to mobile operators (service resellers), who can then combine these wholesales packages to their existing offering and sell the services on subscription basis to the consumers. Service provisioning, marketing, and customer care toward consumers are coordinated by mobile operators. Service reseller and wholesaler have revenue sharing agreements between them.

Most probably the offering to consumers is based on different channel packages of six to ten channels for different user segments and is based on a subscription fee.

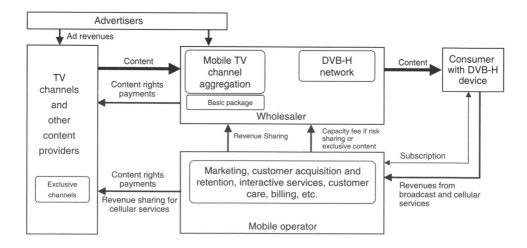

Figure 3.6 Wholesaler–reseller model

In many cases there will be a basic package of mobile TV channel and on top of that the consumer can subscribe premium services and channels. Some of these services can be free to air, funded by advertising or by interactive services.

From the mobile operator's perspective, in this model the DVB-H network is outsourced as is the content aggregation. In the simplest form the mobile operators are just offering a direct customer relationship to the wholesalers in return for a revenue share. Mobile operators are still in control as they cannot be forced to sell the services to their customers.

A variation of the model is where the mobile operator buys part of the network capacity. Then the model is actually very close to what a virtual network operator model is. The difference is that part of the service portfolio (e.g., basic service package) is aggregated by the wholesaler and not by the mobile operator. The mobile operator makes its own content agreements for the exclusive channels. They can use it to differentiate their service offering from the competitors. The mobile operator can acquire, e.g., its own foreign or domestic content and customize the offering to its own customers. In this case, the roles and responsibilities of the mobile operator are the same as in a virtual network operator model. For the broadcast network operators selling the capacity directly to mobile operators offer a way to reduce the financial risk (fixed fee for the capacity).

Often the Mediaset–Vodafone–TIM case in Italy is considered to be an example of a wholesaler–reseller model. Mediaset has built a DVB-H network and is offering the capacity to both Vodafone and TIM. Mediaset is also offering TV channels to the operators as it owns some of Italy's most popular TV channels. Both operators have also exclusive content agreements, e.g., SKY is providing content for Vodafone.

3.6   NewCo or Consortium – Broadcaster-Driven Model

In some markets a special media license is needed for mobile TV channel aggregation. In some countries, only those companies who already have some sort of media or broadcasting license are allowed to either offer mobile TV services or aggregate a mobile TV service portfolio. That might turn into a new type of business model, where the mobile TV channel aggregator can be a totally new company, NewCo, or joint venture or a consortium. It can of course be part of or owned by some of the players in the value chain. The broadcasters or other media companies are most often involved in this model, as they have a good change of winning the media license.

There are several variations of the model. It allows broadcasters to offer services directly to end users, without mobile operators. In this model, the broadcasters (or NewCo) are acquiring capacity from the network operator and basically offering their services to the consumer. This model is the most likely one if mobile TV services are offered free to air, but the model naturally allows also pay TV services.

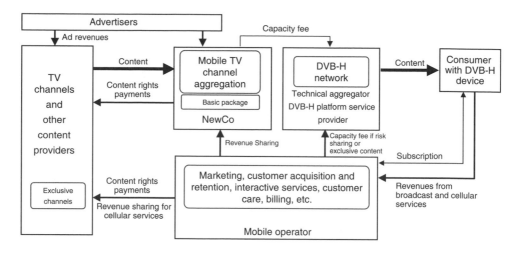

Figure 3.7   NewCo – broadcaster-driven model, mobile operators as service resellers

Another variation of the model is similar to the wholesaler–reseller model. Figure 3.7 illustrates this.

In this model mobile operators are service resellers for the broadcasters, NewCo or consortium. NewCo aggregates the service portfolio and pays for the DVB-H network capacity. Mobile operators take care of the marketing, customer acquisition, customer care, subscription management (delivery of DRM keys), and billing. In return mobile operators have a revenue share with NewCo. Involvement of the mobile operators may vary from the pure subcontractor role (CRM provider) to active channel partner for the NewCo. This is of course reflected in the revenue sharing agreement.

In some countries, depending on the license terms, mobile operators are allowed to purchase part a of the DVB-H network capacity and aggregate their own exclusive mobile TV channels which they offer only to their customers. But this seems to be a rare exception as often in these market mobile TV aggregation is limited only to holder of media or broadcasting license.

A real-life example of the NewCo model is the Mobile 3.0 case from Germany. In Germany local regulators awarded media license to a start-up called Mobile 3.0 (backed by some of the biggest media companies in Germany, and Naspers, South-African media conglomerate). The consortium of mobile operators (Vodafone, T-Mobile and O2) also applied for the media license. Frequency (network) license was awarded for Media Broadcast. Mobile operators and Mobile 3.0 did not find a cooperative model and therefore mobile operators decided not to sell the service of Mobile 3.0 to their customer. This had an extremely negative impact on Mobile 3.0 and eventually forced them to give back the media license in November 2008. The case proved that in a market, where the mobile operators are very powerful and control mobile business, it is very difficult to launch services without their involvement. As of end of 2008, a German regulator is planning to reopen the media license process.

In France, the media regulator CSA awarded a mobile TV media license for 16 content editors, channel providers (BFM TV, Canal Plus, Direct 8, EuropaCorp TV, Eurosport, I-Tele, M6, NRJ 12, NT1, Orange Sports, TF1, Virgin 17 and W9). Three channels are reserved for public broadcasters (France 2, France 3, and Arte). These companies must form a consortium and set up a new company, a special multiplex operator. As of end of 2008, the consortium was still in negotiations with the mobile operators about their role. The key negotiation points are related to who should fund the DVB-H network, mobile operators, or broadcasters, and what is the revenue sharing between these two camps.

3.7 Revenue Models

All of the business models mentioned in the previous sections can have several different revenue models. Revenue models are normally not tied to any specific business model. Revenue models are more likely to be influenced by a service provider's overall strategy and local market conditions.

Revenues in mobile TV business are typically divided into two categories: revenues from the services that are free for the consumers but funded by advertising or by other means and those that consumers need to pay for. Regardless of the source, these revenues are then divided amongst the value chain to cover related investments and operative costs.

For the pure DVB-H network operator, the most common revenue model is a fixed fee for the network capacity. The broadcast network operators often have a monopoly in a country, and therefore the regulator may set limitation on their capacity pricing, to cut off abnormal profits. Normally the pricing must be fair and transparent and the broadcast network operator must offer the same price to everyone. At the same time regulators may have set in the license terms obligations for a certain network roll-out schedule. This is why broadcast network operators typically try to have long-term agreements for fixed prices with the service providers, and are not often willing to accept pure subscribers-based revenue sharing agreements. Of course there are exceptions to that, as some broadcast network operators are willing to take more risk.

From the consumers' perspective, there are a couple of different scenarios for the free services. First is the one where services are offered clear to air. This means no encryption at all; everyone who has a DVB-H enabled device can watch the service. The revenue model is either advertising or promotional (to get the consumer use the service). In the case of public service channels, the revenue model is less clear as in some cases the service provider must offer public channels to the consumers without any compensation and in some case public service channels pay the access fee either for service provider or they directly buy capacity from the broadcast network operator.

The other form of free service is called free to air. In free to air, the service in encrypted, so only those who have subscribed to the service are allowed to watch, even if it does not cost anything to the consumer. This is typically used by the mobile operators, as they often offer part of the mobile TV channels for free, but only to their own customers. The revenue models can vary a lot. Some operators use advertising to fund the channels and some related interactive services. Interactive services, like voting, are of course paid by the users, but mobile TV can still be free of charge. Other operators use free-to-air channels to promote the service, and to attract more people to use the service and maybe even to subscribe to a premium pay TV service. Some use free services to gain a market share, by getting people to change to the mobile operator who offers best mobile TV services. Free to air channel offers are often combined with other cellular service offerings. For example, ten mobile TV channels for free, 500 SMS and 500 min of voice per month for X Euros. In some cases, broadcasters may pay for the service provider to have their mobile TV channel as part of the operator's service portfolio.

For the broadcasters and content owners, free of charge services offer two clear revenue models. Revenues may come from advertising, or from selling the content rights to service providers or from both. Service providers need to buy the content rights for the mobile TV channels they offer to the consumers even if those would be offered free of charge. Content rights are typically different for free of charge and pay TV services; this is why some mobile operators bundle the free-to-air channels with cellular services as explained earlier, and it is clearly stated in their marketing messages that mobile TV is free (consumers may pay monthly fee for subsidized phone, voice and SMS package, but TV is free).

In addition to free services, many service providers offer pay TV channels. Individual events (like a soccer match), TV channels, and channel bundles are sold to the subscribers. In some markets, all the channels on offer are pay TV channels, so consumers cannot watch anything without subscribing the service. Services are typically charged on a monthly basis. Similarly as with free of charge services, many mobile operators bundle mobile TV services with cellular offerings.

A mobile broadcast TV combines best of two solutions: broadcast and cellular. Therefore another logical source of revenue is interactive services. Mobile phones offer a great platform for interactive TV as they are always connected via cellular networks. Mobile operators and other mobile TV service providers use interactive services to enhance the user experience and of course to earn more revenue. Examples for these services are quizzes, voting, and interactive advertising. Also web pages related to mobile TV programs are often considered as interactive services even if there might not be much "interactivity" as such. A service provider could for example offer a mobile TV news channel with interactive links as additional information to the web. User clicks the link and is directed over the cellular network to a related web page. In this case mobile TV has triggered the usage of cellular networks, and thus generates new revenue for the mobile operator. Similarly, the operators can offer a mobile TV program related to premium SMS and MMS services. A typical use case is a music channel, where while watching their favorite artist, the user can buy ringtones, wallpapers, MP3s and videos, and all are delivered over the cellular networks, just one click away. The reason why mobile operators may prefer interactive services is that the revenue sharing can differ significantly from the agreements they have for the broadcast mobile TV services. Mobile operators typically do not share the revenues from additional web usage (triggered by mobile TV) with third parties. Also, e.g., premium SMS and MMS pricing often follows mobile operators' normal revenue sharing in cellular services. Another reason is that mobile operators control the interactivity domain (by owning the cellular networks), which may not always be the case in the overall mobile TV value chain.

There is no clear rule of thumb how the revenues are shared between different players in the value chain, as sources of revenue may differ from one business model to another. The same applies for the revenues sharing. Also operator overall strategies may have an influence on how they prefer the revenues to be shared.

Typically broadcast network operators like to have a fixed fee for the capacity, which means that regardless of how much revenue are generated by the business, they want to earn the same amount of money from service providers for the same capacity and network coverage. In the very early phase of the business this can mean that broadcast network operator gets more than 100% of total revenues, but as business mature and more people subscribe to the services the broadcast operators share starts to rapidly decrease. Of course there are exceptions to this, as some broadcast network operators may have price steps related to number of subscribers, but typically they are risk averse and the pricing is strictly regulated due to monopoly.

Also for the service providers it is not always exactly clear which share of revenues they will get and how that will change over time. Service providers' cost structure is heavily depended on the mobile TV channel line-up and on related content rights. Some channels are more expensive than others, some may be free, and some are even paying for the service provider to be part of the service portfolio. Content rights are also very much dependent on the market and whether the rights are for free services or for pay TV. Good average for West-Central European and US markets for "basic channel" is around 10–20 Euro cents per active mobile TV subscriber per month per channel and for more premium or exclusive channel around 20–35 Euro cents per month per active mobile TV subscriber. This would for example result for 20 TV channel portfolio, for content rights cost of 2–7 Euros per month per active mobile TV user. This money is of course revenues for the content owners and TV channels. So it is extremely important for the service providers to create the optimal mobile TV channel line-up, by thinking of the number of channels (Sky effect) and the balance between basic and premium channels or the split between free adverting funded channels versus pay TV channels etc.

In addition to content rights, the service provider needs to pay for the broadcast network, but also for the marketing, customer acquisition and retention, billing, and other customer care. Very often the mobile operator is the service provider, but in case they are separate companies or if for the book-keeping purposes costs must be separated, service providers need to compensate mobile operators for their services. This cost is dependent on what kind of the role the mobile operator will take. It can vary from only very basic billing (and offering related DRM key distribution over cellular network) to very aggressive customer acquisition campaigns and full customer care and operator heavily involved in service creation. This is again dependent on the local market conditions, but then the average revenue sharing can be between 5 and 50%.

There are many sources of revenues in the mobile TV business, from advertising to pay TV to interactive services. Many operators also use mobile TV as part of their overall service offering, bundling the services with cellular services; this way they can gain more market share, increase ARPU, or reduce churn. Strategies vary from operator to operator. Revenue sharing in mobile TV business depends on business model, operator strategies, what role each company is willing or able to take from the value chain, market conditions and regulation, offered service portfolios and related content rights, and most of all, the revenue shares change over time as business evolves.

3.8   Broadcast Network as Starting Point

In the previous sections we discussed the mobile TV value chain, possible players, and different business and revenue models. But regardless of what role different players want to take in the value chain, everyone should have an understanding of the network investment and the related operating costs and what are the main influencing factors. It is not just pure broadcast network operators who need to understand this, but also all service providers (mobile operators and broadcasters) so that they can evaluate whether the network rent they pay for the broadcast network operator is in the right level, or is the business case better if they build their own broadcast network.

Regardless of the business model or what kind of role different players want or can take in the business eco-system, someone always needs to build (invest into) the network. The key factors influencing the network investment are mainly related to area and terrain types, available infrastructure, available frequencies and generic business drivers like network roll-out, indoor coverage and needed capacity. Other chapters in this book will go into technical details about the influencing factors.

It all starts from the available frequencies. Based on that, the broadcast network operator can start to plan the network design. The operator may have frequencies available for nationwide area coverage or just part of the country, and it needs to agree on the roll-out plan with the service providers, taking of course into consideration the possible license terms set by the local regulator. The roll-out plan may include just some key cities and maybe the connecting roads between them or much more extensive area coverage including most or all of the cities and large rural areas in between those. This naturally has a big impact on the network investment. Typically it makes very little sense to build broadcast coverage into areas where there are only limited number of people, as broadcast networks are more cost-efficient in high population density areas, but it may make sense if for example building the rural and road coverage between two near by major cities creates one larger service area that is more easily understandable and marketable for the consumers. Also in some cases regulator set rules in the network license term for certain roll-out.

Available frequencies also have an impact on building penetration. Lower the frequency, easier the network signal penetrates into the buildings, thus influencing the network investment. Operators cannot of course influence what kind of cities and other residential areas there are in the country. It is very different to build coverage (outdoor or indoor) in central business district, typically packed with huge skyscrapers than in sub-urban residential area with low-rise dwellings with thin walls. Same applies for many "old town city centers" in European countries and Middle East, where building can have very thick walls and the streets are very narrow. In some countries people live more densely than others, some cities are relatively small in area size but people live densely in medium- to high-rise buildings (8–15 floors), and in the other cases cities are huge with hundreds or thousands of square kilometers of sub-urban residential areas. Similarly, the terrain types have an impact on the network, as hilly or mountainous area can be more challenging to cover in some cases than flat terrain, but on the other hand mountains often offer good locations for high power transmitters. All this has a big impact on network investment and efficiency.

Available infrastructure has a huge impact on network investment. In some cities, there are TV towers (200–400 m tall) in the city center. Operators can use high power transmitters (tens of kilowatts) to cover large areas with just one transmitter and maybe some small repeaters or gap-fillers. Other solution is to utilize lower antenna locations like cellular towers or roof tops. In this solution, the same coverage requires larger number of low-power transmitters. Both are good ways to build the network, but the costs may differ significantly from each other depending on the market and company involved. Broadcast network operators typically have an access to TV towers or they even own those themselves; therefore, it is an easy option for them to start building the coverage. Similarly mobile operators may own hundreds or thousands of cellular sites in area where they want to build the DVB-H network, so it is easier for them to start building the network on top of their existing network infrastructure. Site acquisition and rental costs can be quite high if whoever is building the network does not currently own the sites; therefore, the existing infrastructure plays important role, and often affects network design. In the most expensive option, the broadcast network operator needs to build all sites from the scratch. Operators should also consider long-term OPEX cost and not just CAPEX. Therefore very often the network is a mix of high and low power sites, some owned and some rental sites. The typical rule of thumb for the transmitter sites is "higher, the better," meaning that operators should try get access, on average in a covered area, to as high (elevation from the ground level) as possible sites, as this normally reduces both investments and operative costs.

"Higher, the better" is a good rule, but sometimes the proximity is more important. In some cases TV towers are not optimally located in city centers, or close to where people live, but outside of the city. When the distance to city center or the main urban areas increases, then broadcast network operator needs to increase transmitter power heavily to cover the area. There might be limitations in frequency design, which does not allow too high transmitted powers (e.g., because of the neighboring countries, or Single Frequency Network design), or it is simply less expensive (both CAPEX and OPEX or longer term optimal combination of those) to build coverage with small power transmitters, which are located in the proximity of city center or urban areas.

Service providers need to decide how many mobile TV channels and other services they want to offer to their customers. This has a direct influence on the needed capacity. For example, 15 mobile TV channels with 300 kbps require roughly 5 Mbits of network capacity (4.5 Mbits for TV channels, some hundred kilobits for encryption key streams and around 300 kbps for ESG). For 30 channels capacity requirement naturally doubles if the channel quality remains the same. The main modulation options operator can choose from are QPSK, which roughly gives 5 Mbits of capacity, and 16-QAM, with roughly 10 Mbits of capacity. QPKS modulation is quite robust modulation and therefore with that it is easier and less expensive to build coverage than with 16-QAM modulation. This means that fewer transmitters are needed in the QPSK network than the 16-QAM network. Drawback is of course that QPSK offers less capacity than 16-QAM. The rough rule of thumb is that the QPSK network CAPEX is around a half of 16-QAM network investment. This is of course dependent on local market conditions, area to be covered, and especially what kind existing infrastructure is used.

Another very influencing factor is required outdoor and indoor coverage. This is again a factor that service providers need to decide, as they need to explain to the consumers what the service area is. As mobile TV is used mainly with mobile phones, consumer might expect similar coverage for mobile TV as their current cellular coverage is. Therefore, it is very important for the service providers to know what kind of indoor and outdoor coverage is needed. It is not enough just to agree with the broadcast network operator about some area coverage; they also have to go into the details for the best possible solution for the consumers. Especially mobile operators and broadcast network operators may have a very different view on what is good coverage. Indoor coverage is much more difficult to achieve than outdoor coverage and it of course also depends on area type. Buildings are different in different areas (Skyscrapers district vs. sub-urban residential area), which influences how difficult it is to achieve good indoor coverage. Basically good indoor coverage requires more transmitters and more transmitted power, thus increasing investments and operative costs compared to outdoor only network. The rough rule of thumb is that investment in outdoor coverage is around a half or one third of good indoor coverage investment.

As this book focuses mainly on technical aspects of the broadcast networks, the other areas like marketing, customer acquisition cost, content rights, mobile TV ARPU and subscribers' forecasts, and also the detailed business cases for different players, including broadcast network operator, in different business models remain out of the scope of this study. The following chapters will go into deep technical details about the broadcast network.

4

Technical Architecture

The DVB-H network covers the entire network infrastructure within the DVB-H network. The whole chain from the content delivery to the service systems, until to the broadcasting of the DVB-H signals is the task of the DVB-H head-end.

DVB-H standard specifies the network functionality for the radio interface, including the relevant parameters. DVB-H is thus the access network for the Mobile Broadcast system behind the radio, delivering the IP packets from the source up to the IP Encapsulator (IPE) that is located in the border of the DVB-H network. IP is used in connections between the content provider, distribution network and the terminal. The Broadcast Encapsulator, i.e. IPE Manager and a set of IPEs, functions as an interface between IP and DVB-H networks.

4.1 Main Elements

As the core network of DVB-H is based on packet data transmission, the elements are connected to each other via the IP network. The capacity of the distribution network should be thus dimensioned sufficiently high in order to avoid possible bottlenecks in the transmission.

The complete end-to-end DVB-H service chain consists of the DVB-H domain, which delivers the signal for the DVB-H terminal, and the IP network domain, which transports the contents from the encoder to the DVB-H domain. Figure 4.1 clarifies the division.

As can be noted for Figure 4.1, the IP encapsulator (IPE) acts as an interface between the IP networks and DVB-H broadcast network. The DVB-H standards

The DVB-H Handbook   Jyrki T.J. Penttinen, Erkki Aaltonen, Jani Väre and Petri Jolma
© 2009 John Wiley & Sons, Ltd

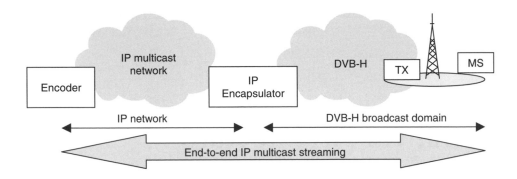

Figure 4.1 The end-to-end DVB-H service consists of the IP network and the actual DVB-H broadcast network

define the latter one, leaving the practical solution designing of the IP multicasting for the related network element and service vendors.

DVB-H has been designed in such way that it is backwards compatible with the digital terrestrial broadcast network DVB-T. This means in practice that the contents of both DVB-T and DVB-H can be offered via the same physical site and radio bandwidth by multiplexing the contents of each one.

Each DVB-H stream, i.e. transport stream (TS), is multiplexed for the IP multicast network domain. The basic principle of this can be seen in Figure 4.2.

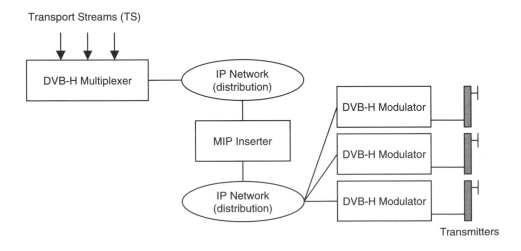

Figure 4.2 High-level DVB-H architecture. The source streams are multiplexed and distributed over the DVB-H IP network, and delivered to the terminals via DVB-H transmitters. If the Single Frequency Network is applied, the correct synchronization of the transmitters must be taken care of by using a MIP inserter

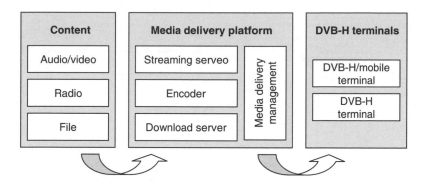

Figure 4.3   The block diagram of the DVB-H architecture

In the practical DVB-H setup, the core network consists of elements (servers) that manage the streams (IPE and its management element) and take care of the customer interactions (authentication, authorisation and billing).

As an example, Figure 4.3 shows the high-level DVB-H architectural solution. The architecture can be divided roughly into four parts. The content is the original source of the streams e.g. video and/or audio programs or data files. The content streams are converted as IP packet streams, and handled by the media delivery platform. As a third part, there is infrastructure for the supporting functions such as charging, operations and maintenance and customer care. The final part of the architecture consists of the DVB-H devices which can be either stand-alone or integrated with 2G and/or 3G terminals. The former provides with the means of using interaction channels via e.g. GPRS.

Figure 4.4 shows a more in-depth view to the DVB-H network with the respective elements. The division can be made between the actual DVB-H network and the supporting mobile communication network. As can be seen from the figure, the source contents is handled by a set of encoders, i.e. each encoder handles a single program. The encoder pool can be set up in a static or dynamic way, i.e. the encoder can use either fixed bit rates (CBR, Constant Bit Rate) or VBR (Variable Bit Rate). Depending on the equipment and capacity requirements of each captured program, the used bandwidth may also vary in a dynamic way balancing the load between the encoders. This provides with enhanced quality with the same total bandwidth of the core network as the effect is the same as in fixed and mobile telecommunication networks behaving according to the Erlang B model.

DVB-H consists of both DVB-T and new DVB-H-specific functionalities. The idea of the main functionality of the DVB-H network is presented in Figure 4.5. As the convolutional coding of the radio interface is the same for DVB-T and DVB-H, the system is backwards compatible providing the possibility of multiplexing both DVB-T and DVB-H streams in the same radio frequency band.

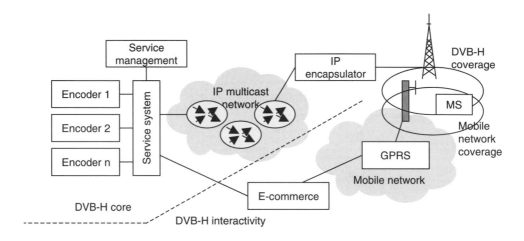

Figure 4.4   Main elements of the DVB-H network

In the receiving end of Figure 4.5, the DVB-H receiver consists of the demodulator and terminal. The DVB-H demodulator gets the DVB-T signal via the RF input either from the internal or optional external antenna. The DVB-H demodulator block is in fact a DVB-T demodulator with the DVB-H-specific 4K and TPS functionality added on it. The demodulator block also contains time slicing with power control functionality as well as optional MPE-FEC. The DVB-H terminal part obtains the streaming data (IP datagrams and TS data) from the demodulator block.

4.2   Core Network

The following sub-sections provide an overview of the main elements within the core DVB-H network. The more specific description of the each element can be

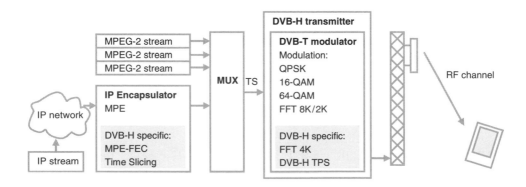

Figure 4.5   It is possible to multiplex the DVB-H IP streams with the DVB-T MPEG-streams. The streams can be delivered via the common infrastructure for the DVB-H and DVB-T terminals

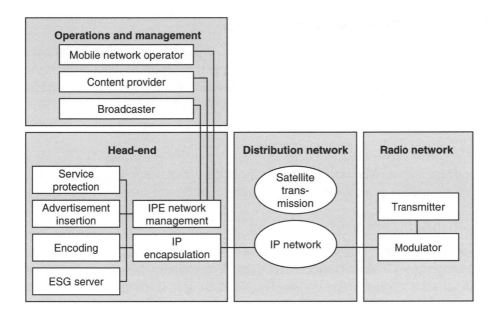

Figure 4.6   The layout of the delivery chain within the core DVB-H network

found in Chapter 6. Figure 4.6 illustrates the generic layout of the delivery chain within the core DVB-H network, where the main elements are *Operations and management, Head-end, Distribution network and Radio network.*

Next, Figure 4.7 presents an example of the DVB-H network design, which is based on the Nokia MBS 3.2 Mobile Broadcast Solution. As can be observed from Figure 4.7, the broadcast network is unidirectional, sending data streams in downlink. The possible return channel can be done by using already existing mobile communication networks, GPRS being the most logical solution for the interactions.

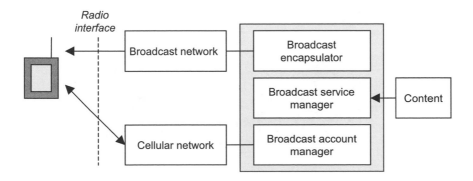

Figure 4.7   An example of the DVB-H network design according to the Nokia MBS 3.2 Mobile Broadcast Solution

4.3   Mobile Broadcast Solution (MBS)

The MBS is an open standard server solution to support broadcasting digital content via DVB-H to mobile terminals. It can utilize e.g. the current TV content as an input, as well as data files arranged in repeating carousel. It consists of BSM, BAM and encapsulators as shown in Figure 4.8.

4.3.1   BSM

Broadcast Service Manager (BSM) controls the encapsulation, multicast routing, encryption, ESG generation and DRM aspects of mobile broadcast services.

4.3.2   BAM

Broadcast Account Manager (BAM) is an on-line service fulfillment and charging solution for paid mobile broadcast services. A consumer who selects a pay service from the ESG is prompted to try it and buy it. If the user wants to purchase a content, a request is routed via the cellular packet IP services to BAM.

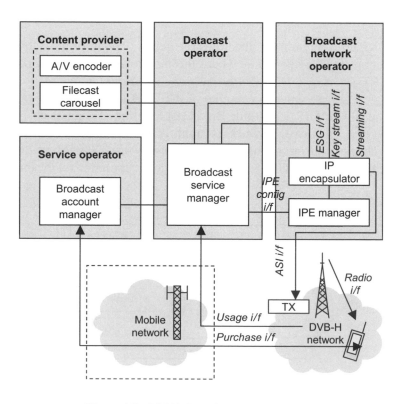

Figure 4.8   Mobile Broadcast Solution (MBS)

4.3.3   Encapsulator

Broadcast Encapsulator (BE) functions as the IP to DVB gateway network with encryption capabilities. The IP forwarding and the encryption settings are centrally managed by the BSM. BE consists of IPE and IPE Manager.

IPE is a gateway between the IP multicast network and the DVB-H transmitters. Each IPE10 routes the applicable multicast groups. BSM controls both the routing and the encryption of each IPE element. IPE encapsulates the multicast datagrams (UDP) of the streamed content into MPEG-2 compliant transport streams.

IPE Manager is a network element enabling central management of all IPE elements in a given network. It mediates the control messages from BSM to each IPE element.

IPE can be considered as a translator between the IP core network and the DVB-H modulator. It is thus located in the edge of the transport (core) and radio networks of DVB-H.

The audio/video and/or possibly other streams from the encoder side to the broadcast encapsulator are delivered via IP multicast. The broadcast encapsulator concept consists of IPE Manager and various IPEs. The basic principle is that the stream is actually shared as long time as it is physically just possible (Figure 4.9). Each

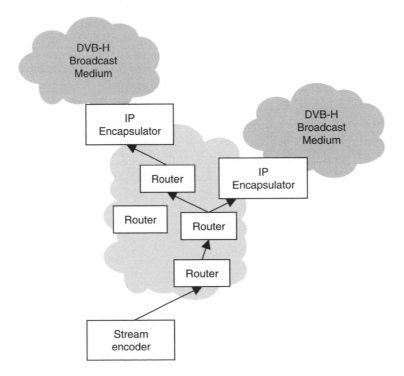

Figure 4.9   The principle of locating IPEs. Both the encoders and routers do have support for the multicasting

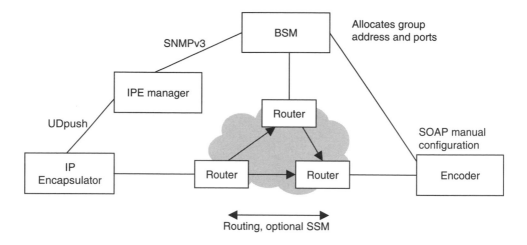

Figure 4.10 IPE registration process

IPE receives the multicast stream from the encoder through an IP network, which can be Intranet, Extranet, or public Internet. The transmission can also be tunnelled between the encoder and IPE. The advantage of using multicast is obvious when multiple IPEs are used as the streams are replicated as close to the IPEs as possible.

The IPE's role is to be a multicast client. It registers itself as a listener for all group addresses the BSM has allocated via control commands (Figure 4.10). The multicast address must be set in the encoder. BSM can signal the multicast group addresses and port ranges via SOAP (Web Services interface).

As for the terminology of the DVB-H, the cell refers to the radio coverage area that is created via the delivered streams within a single IPE. In the Single Frequency Network, multiple physical sites are connected to each IPE, which means that a single cell can thus consist of several transmitter sites. This term differs from the mobile network systems where cell refers normally to the coverage area that is obtained by a single transceiver element with possibly sectorized cells, each site sector or omni-radiating antenna system presenting one cell.

4.3.4 Encoders

The main idea of the encoders is to receive the contents e.g. audio/video streams, and convert them to the format that the DVB-IPCD network routs for the radio network and finally to the terminal.

4.3.5 Network Management System

The network management system depends on the network solution. Various options can be utilized in the operations, including the remote usage of the elements.

In addition to the network management (i.e. change of the parameter values remotely), the network management system can handle the fault management, performance monitoring. Also other management tasks might be possible, i.e. inventory management (in equipment HW and SW level), as well as backup and restore functionality.

4.4 Radio Network

4.4.1 Sites

The DVB-H radio transmitter site consists of a transmitter and modulator. The single site architecture is thus relatively simple. The digital DVB-H transmitter contains the modulator that provides typically power levels of some hundreds up to some thousands of watts. If the repeater solution is used, it can be a stand-alone element receiving and re-sending the signal in the areas that are challenging to cover via the main transmitters. The repeater is often referred to as gap filler in DVB-H terminology. The power level of the repeater is normally from some watts to hundreds of watts.

The DVB-H transmitter can also be of analogue power amplifier. In this case, the DVB-H modulator is connected in front of the power amplifier. It is worth noting that the calculation of the transmitter power level in the analogue amplifier case differs from that in the digital transmitter case.

The other needed elements in the site are GPS and related MIP inserter in order to provide an accurate synchronization (time stamp) for the single frequency network. They are connected to the transmitter in the same site. The GPS signal can be delivered to the transmitter via coaxial antenna cable, but it is worth noting that the cable attenuation might get too high when using the thinnest coaxial cables with long distances e.g. from the rooftop to the transmitter input.

The site also contains other related materials such as antennas, antenna feeders, and possible power splitters for the divided directional antenna elements, tower in field or pole or other mounting mechanisms in the rooftop area. Also the power rectifiers and possible backup system are part of the site equipment.

4.4.2 Transmission

The transmission between the radio transmitter site and the core network (IPE being the interfacing element) can be done physically via terrestrial links (e.g. radio link, fibre optics, etc.) as long as the capacity and quality requirements are fulfilled.

The transmission can also be based on the satellite link. In this case, the site contains an additional satellite transceiver.

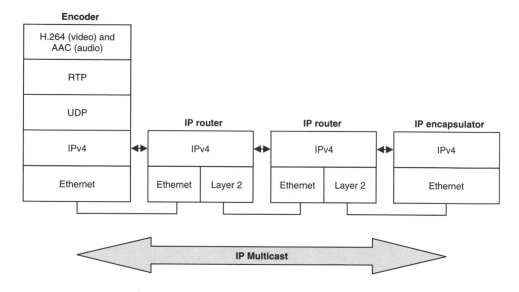

Figure 4.11 The protocol stacks of DVB-H transmission

4.5 Interfaces and Protocols

The DVB-H core network is based on the IP Multicast with related elements participating in the stream delivery. The core network can thus be considered as a router network with the participating servers connected to it, encoder producing the streams to be delivered for the IP encapsulator. Figure 4.11 presents the protocols of the DVB-H core elements.

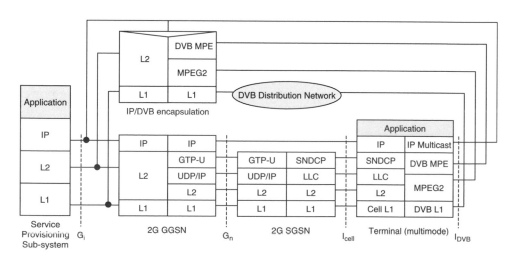

Figure 4.12 The protocol stacks in the complete DVB-H network

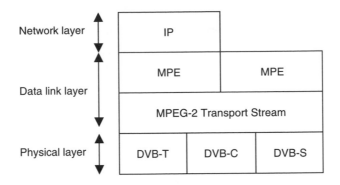

Figure 4.13 The layer structure of DVB-H

The packet transmission between the encoder and IPE consists of IP packets with variable length. The maximum packet size is less than 1400 bytes. The IP Multicast functions in this path, with IP routers participating in the packet delivery. It is worth noting that the IP packet TTL should be set long enough in the encoder.

The IPE is connected to the DVB-H modulator and transmitter via ASI. It is in practice a coaxial cable.

The radio interface is based on the digital modulation (QPSK, 16-QAM and 64-QAM). The DVB-H radio path is only defined in the downlink direction. For the possible interactions, a separate radio system can be used, GSM/UMTS being the most logical as the DVB-H terminal often includes it by default. Figures 4.12 and 4.13 show the DVB-H layers and protocol stacks.

5

DVB-H Equipment

The complete DVB-H solution consists of core network, radio network and terminals. Each element participates in the content delivery chain. There are various vendors for each one of the elements. Some of the elements are DVB-H specific (e.g. terminals), whereas the IP core network is based on the common IT technologies like routers, servers and local area network layouts.

5.1 Terminals

The first prototypes of the DVB-H terminal were published in 2004. The very first DVB-H terminal was the Nokia 7700, which contained a separate DVB-H module that was possible to add on top of the battery container at the back of the phone (Figure 5.1). The equipment was used in the early phase pre-testing e.g. in RTT field trials in Helsinki, Finland, but the model was never launched commercially.

Figure 5.2 shows a high-level block diagram of the DVB-H receiver. The reception of the Transport Stream (TS) is compatible with DVB-T system, and the demodulation is thus done with the same principles also in DVB-H. The additional DVB-H-specific functionality consists of Time Sliced burst handling, MPE-FEC module and the DVB-H de-encapsulation. Figure 5.2 does not highlight the extraction of PSI/SI signalling information, which is also usually done by the DVB-H receiver part.

As can be seen from Figure 5.2, the frame error rate (FER) information, i.e. frame errors before MPE-FEC-specific analysis, is obtained after the Time Slicing process, and the MFER (frame errors after the MPE-FEC correction) is obtained after the MPE-FEC module. If the data after MPE-FEC is free of errors, the

The DVB-H Handbook   Jyrki T.J. Penttinen, Erkki Aaltonen, Jani Väre and Petri Jolma
© 2009 John Wiley & Sons, Ltd

• DVB-H terminals

Figure 5.1   Some DVB-H receiver models from Nokia

respective data frame is de-encapsulated correctly and the IP output stream can be observed without disturbances.

In addition to the DVB-H receiver, the terminal also contains the related functionality to show the DVB-H contents as well as the user interface. The former can consist of commonly available multimedia streamer that can present the video and audio streams by buffering the contents. For the latter one, basically all the methods to input the commands are possible, including traditional key pad, touch screen and voice command interface.

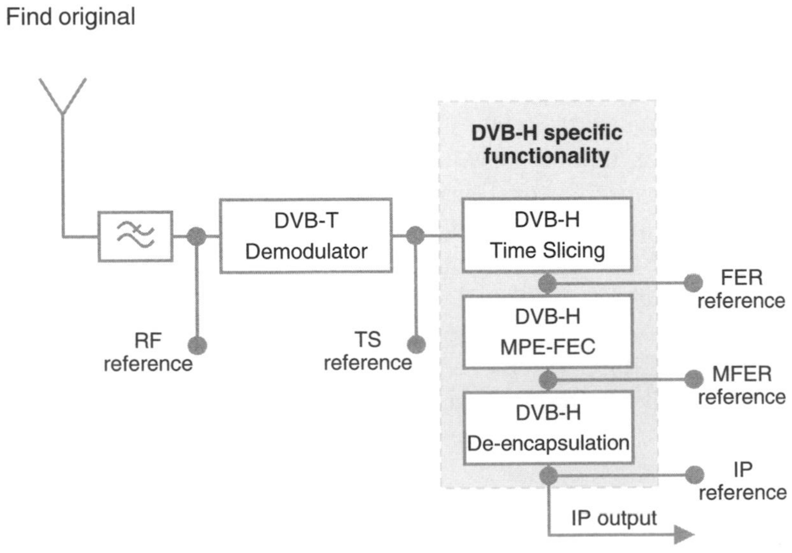

Find original

Figure 5.2   A principle of the reference DVB-H receiver

The most logical DVB-H terminal type consists of DVB-H receiver with the related functionality (video streamer application, etc.), as well as mobile communications network, which can be e.g. GSM and/or UMTS. The simultaneous support of both of these network types eases the interaction processes e.g. the selection of non-free-to-air channels. As the current mobile networks contain by default the AAA procedures, i.e. authentication, authorization and accounting via the subscriber registers and using the subscriber codes (IMSI, MSISDN) and SIM/USIM, the internal connection of the mobile networks and DVB-H core network is straightforward.

In addition to the combined DVB-H receiver and mobile communications terminal, one possibility is to have stand-alone DVB-H terminals for the reception of purely free-to-air DVB-H signals. By applying e.g. scratch cards or remote (e.g. Internet based) subscription and local subscription management right methodologies, it is also possible to observe closed channels with this type of terminal.

The DVB-H terminal also contains the related functionality to show the DVB-H contents as well as the user interface. The former can consist of commonly available multimedia streamer that can present the video and audio streams by buffering the contents. For the latter one, basically all the methods to input the commands are possible, including traditional key pad, touch screen and voice command interface.

The terminal often contains the mobile network terminal with related SIM/USIM in order to ease the interactions.

Mobiles have been divided into different classes based on the functionality and technical possibilities. These four classes are defined by the DVB project and are:

- Class A is meant for the outdoor pedestrian reception. It is a receiver with attached or built-in antenna.
- Class B is meant for the indoor reception. It is a receiver with attached or built-in antenna. There are further two sub-classes:
  - Class B1: Light-indoor reception, receiver close to a window in a lightly shielded room.
  - Class B2: Deep-indoor reception, receiver further away from a window, in a highly shielded room.
- Class C is meant for the mobile reception. It is a receiver in a moving vehicle with external car antenna and different speeds.
- Class D is meant for the mobile reception. It is a receiver in moving vehicle, with attached or built-in antenna and different speeds.

Table 5.1 shows the different mobile classes with the reception capabilities. From this table, it can be seen that all the mobiles can receive UHF IV and UHF V bands.

In the very detailed radio network planning, the differences of the terminal performance could be taken into account by carrying out sufficient amount of

Table 5.1   The DVB-H terminal categories

Terminal category	VHF III	UHF IV	UHF V	Environment
A	Yes, in areas where VHF is in use for DVB-T	Yes	Yes	Integrated car terminals
B1	Yes, in areas where VHF is in use for DVB-T	Yes	Yes	Portable digital TV sets
B2	Optional	Yes	Yes	Pocketable TV sets
C	No	Yes	Yes, up to channel 55	Convergence terminals

laboratory tests in order to have statistically significant understanding about the variations of the performances between different models. In the most accurate case, the planning criteria can be varied in different areas of the network depending on the distribution of the terminal models. In practice, the differences are very challenging to take into account, and a common value for link budget can be used instead in a sufficiently good radio planning.

5.2   Core Elements

A DVB network usually consists of a number of elements. The main function of the elements is to convert MPEG-2 or any other format of video into a transport stream towards the transmitter and to the mobile. Along the way, the signal is being transformed, coded and protected against air interface errors. The main elements are shown in Figure 5.3.

5.2.1   MPEG-2/H-264 Encoders

The encoder's main function is to compress the delivered uncompressed video (SDI) stream into MPEG-2, MPEG-4 or H-264 format. This is usually the first element in the DVB-H network. Inputs are connected 'directly' to the broadcast TV stream output. Encoders are usually supporting different standards like PAL and NTSC, and different audio possibilities.

5.2.2   DVB Modulators

The DVB modulator can be considered as the heart of the DVB-H system. The modulator takes care of the error correction functionalities, mapping, shaping and re-sampling and is the last step before converting and transmitting the signal over the air.

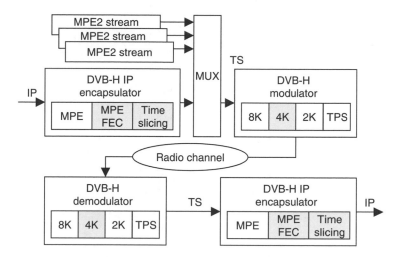

Figure 5.3 The high-level principle of DVB-H transmitter and receiver. The grey colour indicates the additions that DVB-H brings compared to the previous DVB-T based solution

In Figure 5.4, all the functionalities are shown and different vendors might have one or more devices.

The modulator converts the signal to QPSK, 16-QAM or 64 QAM format. As Figure 5.5 shows, more symbols per bit is transmitted, more complex the reception becomes, and the higher data rates thus require accordingly better received power levels.

5.2.3 Transport Stream Multiplexers

The main functionality of the multiplexers is to combine a number of input streams from one or more television providers into one (HP) or two streams (HP and LP) and to add the necessary system information signalling messages.

Depending on the manufacturer the functionalities may vary. Usually the multiplexers are supporting PID remapping, PSI/SI adding and have input possibilities for an external EPG or external server for addition of large quantity of information.

The output of the multiplexer is e.g. DVB-ASI format in DVB-H, and is then ready to be encrypted and error protected.

5.2.4 EPG Generators

The Electronic Program Guide (EPG) is one of the elements which are optional in the network. The EPG has been standardized by ETSI and contains information

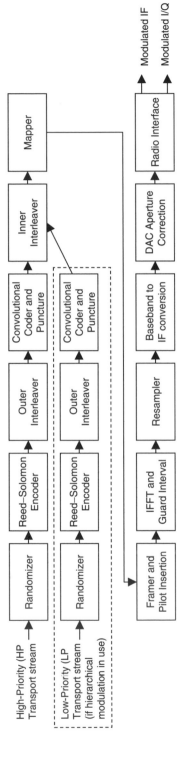

Figure 5.4 The block diagram of DVB-H modulator

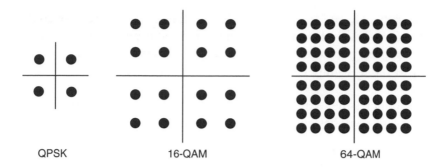

QPSK   16-QAM   64-QAM

Figure 5.5 The constellation principles for the modulation options of DVB-H

regarding the programs that are transmitted at the moment and programs that will be transmitted. EPG is also available in most digital television broadcasts.

5.2.5 SFN Adaptor/MIP Inserter

SFN Adaptor inserts Megaframe Initialization Packet into MPEG-2 TS periodically, and controls all the transmitters in SFN network to emit the same signal bits at the same time. The adapter can configure the MPEG TS stream according to the TPS.

5.2.6 Distribution Network

The distribution network is the connecting element between the core elements and the radio elements. In most of the cases the core elements are grouped together while the radio elements are distributed in the required coverage area. This distribution network can consist of any methodology keeping in mind the input requirements of the transmitter (usually that is an Ethernet connection). The most cost-effective solution, at least in many countries, is an IP network. IP networks are fairly easy to install and can support the required bit rate for the transmitters (up to 32 Mbps). Other options are WiMax, Ethernet, satellite (additional synchronization needed), and microwave links.

5.2.7 Digital-to-Analogue Converter

The last step is the conversion of the DVB-H stream into an analogue signal that can be transmitted over the air.

The final architecture of the DVB-H core network solution depends on the element and core network service providers. DVB-H standards do not specify the more specific realization of the core functionality whilst the source (audio/video/

files) can be transferred from the encoder or other source to be delivered via the radio interface. DVB-H basically specifies the radio path with relevant coding schemes, frequency bandwidths, technical parameters and the signalling that is provided to the transmitter via the IP Encapsulator, and delivered finally to the terminal.

As an example, the Wipro MBS 3.2 (Mobile Broadcast Solution) consists of several servers, switches and DVB-H-specific modules.

5.3   Radio Elements

The radio network requires sufficiently powerful transmitters in order to cover the planned geographical area with selected parameter values.

There are several DVB-H transmitter providers in the market. The maximum power levels of the DVB-H transmitters can be up to about 10 kW, but for the practical reasons, the output power of the power amplifiers are normally maximum of some thousands of watts. More power is provided, the power consumption increases accordingly. Also the complexity of the transmitters increases due to the need for the liquid cooling instead of air cooling. Also the physical size of the transmitter increases, and requirements for the auxiliary power systems become more complex, so there is an obvious optimal limit for the practical power levels.

The DVB-H transmitter consists of connections to the DVB-H antenna system (coaxial cable), GPS antenna (for the reception of the synchronization signal) and ASI connections towards the core network.

DVB-H transmitters can typically be managed via a remote connection, which is based on LAN in practice. Basically all the functions, including parameter settings, performance monitoring (including the transmitted and reflected power levels) and fault management (alarms monitoring, resetting the transmitter in case of the power-down) can be handled via the remote connection without physically visiting the transmitter site unless there is a hardware problem or other severe fault in the equipment or connection.

5.4   Antenna Systems

DVB-H antenna system consists of the antenna cables between the transmitter and antenna elements, jumpers, connectors and possible power splitters, as well as the mounting material like brackets and antenna towers or poles. Depending on the location, power levels, etc. the antenna elements can also be installed on rooftop, inner or outer walls and indoor roofs.

The suitable antenna elements for DVB-H can be the same type than are used in typical broadcast solutions, although there are specifically designed antennas for the DVB-H due to its lower power levels. In order to select the correct antenna type, the

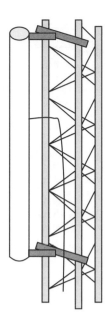

Figure 5.6 An example of the antenna installation on the tower

network designing in short and longer term should be known in order to have the
correct power constraints, etc. noted. This prevents the need to change the elements
in the later phase e.g. due to the raised transmitter powers.

The DVB-H antennas can be categorized basically as outdoor and indoor
antennas. Outdoor antenna can be either omni or directional (omni radiation
pattern can also be created in practical by combining several directional
antennas), whereas the indoor antennas are of low power types, normally
omni-directional elements in order to cover isolated areas as shopping centres,
airports etc.

Omni antennas are in general a few metres high and have a gain of around
8–10 dBi. To get more gain omni antennas can be stacked (see Figure 5.6).

More commonly are the quasi-omni antennas. These antennas consist of 1 to 18
panels as shown in Figure 5.7, depending on the required radiation pattern and can
be stacked together to obtain higher gains. The most used configuration is with four
panels.

DVB-H antennas consist of several elements, or panels. In Figure 5.8, configurations with respectively two, three and four panels have been shown. Depending
on the required gain of the antenna, a number of panels can be stacked on top of each
others and can form rather big installations up to about 16 m of height. The number
of antennas in stacked configuration dictates the bays, i.e. the number of bays
represents the number of stacked antennas.

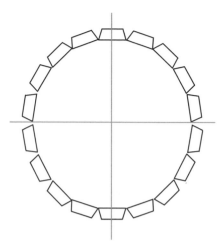

Figure 5.7   An example of the omni-radiating site cell by using various separate directional antennas

The directional panel antennas are easier solution for major cases for the DVB-H site installations (Figure 5.9). The directional elements can be installed e.g. in the middle of the tower one-by-one, in each side, in order to create the omni-directional pattern. The additional advantage is that the elements can be down-tilted separately in order to create an optimal coverage area in each side of the tower. The single elements are more feasible to transport and install, and the achieved antenna gain can be considerably higher compared to the pole element solution. The directional panel antennas can also be installed relatively easily to walls and rooftops in such a way that the back lobe does not provide excessive amount of radiation power inside the building or within the rooftop. This controlled radiation power limitation makes the maintenance work easier, and it is thus not needed to switch off the DVB-H transmitter when other personnel climb onto the rooftop.

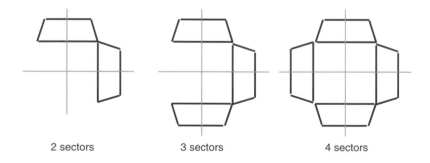

2 sectors   3 sectors   4 sectors

Figure 5.8   Three examples of the directional antenna positioning

Figure 5.9 An example of a light-weighted DVB-H antenna setup with portable pole base. This case consists of two directional elements on top of each others. The setup with variable number of antenna panels is suitable e.g. for the initial testing of the radio network

5.5 Measurement Equipment

The DVB-H network consists of various elements and interfaces, which require both fault management and performance monitoring in the operational phase of the network. The possible faults can be originated e.g. from the service protection state, encoder status, the presence of the ESG, IP encapsulator status, services status, the functioning of the distribution network (jitters and faults), the functioning of the radio network (transmitters status) and the functioning of the modulator.

There is a variety of measurement equipment available for the DVB-H testing both for the core network side as well as for the radio interface. In addition to the need for the performance monitoring, the DVB-H network consists of several potential fault causes.

In the next sections, both the core measurement equipment and the radio measurement equipment are going to be described.

5.5.1 Core System Evaluation

In the core side, normal IP-related measurement and analysis equipment can be used. As an example, an MPEG stream monitor might be capable for multi-layer, multi-channel and remote monitoring and measurement at IP, RF and transport layers to DVB-H. The key IP parameters can also be monitored and simultaneous monitoring of IP and MPEG layers provide more detailed and easy fault management and performance monitoring. The equipment might include UDP, RTP, with Internet Group Management Protocol (IGMP), Address Resolution Protocol (ARP) and Internet Control Message Protocol (ICMP Remote ping) for the detailed analysis.

5.5.2   Radio Interface Evaluation

In the radio interface, basically two different types of measurements can be done.

1. 'General' measurements – this means that the measurements are conducted in such a way that the basic signal level is recorded and/or shown together with the selected error indication (e.g. BER, FER or MFER).
2. 'Deep analysis' measurements – in this type of measurement a closer look is taken into the distribution of the OFDM signal together with an error cause and their distribution.

5.5.2.1   General Measurements

Figure 5.10 identifies different points of measurement. These process points are representing different stages in the decoding and presenting the IP stream on the mobile. Each point can be used in verifying certain aspects of the network.

The different measurements which are usually supported by different measurement vendors are as follows:

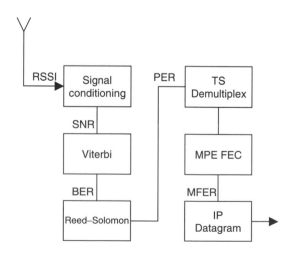

Figure 5.10   Extracted presentation of the measurement points as interpreted from ETSI TR 102 611. RSSI refers to Received Signal Strength Indication, SNR to Signal-to-Noise Ratio, PER to Packet Error Ratio and MFER to MPE-FEC Frame Error Ratio

Received Signal Strength Indicator (RSSI)
The RSSI shows the level of the received signal. Usually RSSI indicates the total received power level at the mobile. This means that the value is showing the best received signal at a particular time and place and it does not indicate where the signal is originating from (serving or neighbouring cells).

Derived Power (DP)
When the signal is processed by the receiver, signal-conditioning will be performed. In this signal-conditioning stage, there will be a number of filtering and gain stages which have as a result the possibility to derive a power level for the received 'wanted signal'. This derived power level may be known as the derived power, and it is a good early indicator of the possible quality-of-signal available. The derived power does not distinguish between the signal power and the noise power.

Signal-to-Noise Ratio (SNR)
The signal-to-noise ratio is the first indicator for the quality of the received signal and is measured before Viterbi decoding. This parameter is used in analogue and digital networks. The value represents the coverage at a certain point in space and time. A lower SNR indicates that the carrier signal is close to the noise floor. The minimum SNR is depended on the network settings. The actual figures will be presented in Chapter 9.

Sometimes this value is also expressed as the carrier to interference plus noise ratio $C/(I + N)$, which basically indicates the same, with the only difference that the figure shown can be either for the interferer or noise. This SNR is an early indication for the quality of the level.

Bit Error Ratio (BER) before Viterbi
The BER is the digital SNR and represents the probability of having errors and can be measured before (BER) and after Viterbi (VBER) decoding. Bit error can be caused by different aspects in the network. Main contributors are terminal velocity, interference and coverage (RSSI). The likelihood of an increased BER due to velocity is in general not applicable. Cases where the terminal velocity plays a role are at the moment limited to high speed trains (300+ km/h). The main contributors are either interference, this could be seen with an increased $C/(I + N)$ or coverage which should have a lower RSSI. The BER itself does not provide sufficient information to proceed further.

The BER value should be in general lower the 1 bit error in 10,000 bits (1×10^{-4}). DVB-H Implementation Guidelines indicates that for DVB-H, a QEF criteria of $2 \times (10^{-4})$ can be used in order to identify the practical quality limits of the signal. The BER is a good indicator for the system performance on an independent error on a bearer, but it has little meaning on bursty or depended error channels.

When the BER exceeds a critical value, and the RSSI is low, the mobile will start to tune to a different channel which is available, if there is one. This means that the BER is an indicator to network coverage planning.

Bit Error Ratio after Viterbi (VBER)

The Viterbi error correction is based on Quasi Error Free (QEF) quality target and should be less than one uncorrected error per transmission hour, corresponding to a bit error ratio of 1×10^{-11} when the RSSI and $C/(I + N)$ criteria are fulfilled.

The VBER decoding can be interpreted in the same way as the Bit Error Rate before Viterbi. The DVB-H Implementation Guidelines informs that for the functional QEF point of the VBER 2×10^{-4} value has been indicated to reflect the reality, but VBER is not very useful as such, and should thus always be checked with other values like the $C/(I + N)$ and the RSSI. This is due to the fact that in the hardly functioning area of the system the terminal might not interpret accurately the VBER due to the excess off errors.

Packet Error Ratio (PER)

The PER can be used for slowly degrading signal quality and is generally used by the mobile as an indication that other signals should be monitored. PER is measured as the non-correctable RS packets that were received. The criteria of the PER have been set on 1×10^{-4}. The PER indicates the number of erroneous errors in the video on the mobile screen before FEC correction.

The measurement period should be at least 800.000 TS packets (about 2 min with 16-QAM, CR = 1/2, and GI = $^1/_4$).

Frame Error Rate (FER) before FEC

Frame Error rate before Forward Error Correction (FEC) is an indication of the Quality of Service of DVB-H, since it shows the number of erroneous frames out of the total number of received frame before the error correction. In networks without MPE-FEC this is the first suitable criteria to evaluate the quality of service.

A frame is marked as erroneous if at least one TS packet within a frame is erroneous. The quality of service point has been set to 5%.

Frame Error Ratio after FEC (MFER)

The MFER indicates the slow degrading signal quality and is referring to the error ratio of the time sliced burst protected with the MPE-FEC. This MFER is sometimes considered as the first suitable degradation criterion (DVB-H implementation guide – A092 DVB Bluebook). Frame Error Ratio after FEC informs the number of erroneous non-recoverable frames out of the total received frames. In DVB-H, the MFER criteria have been set to 5% (same as for the FER) and indicates the

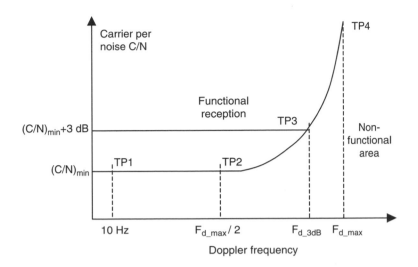

Figure 5.11   A principle of the behaviour of the MFER as a function of the terminal speed (Doppler shift)

degradation point of the DVB-H service.

$$\text{MFER}(\%) = \frac{\text{Number of Erroneous Frames} \times 100}{\text{Total Number of Frames}}$$

In Figure 5.11, the MFER of 5% is represented by the fixed line. The area which is marked as working reception is within the 5% MFER. The C/N_{min} is the minimum required C/N value and the Doppler frequency represents the terminal velocity expressed in hertz. A minimum of at least 100 frames should be analysed in order to have a valid measurement.

5.5.2.2   Deep Analysis – Constellation Errors

In the deep analysis a closer look is taken into the signal usually in bad radio condition areas, like where there are a lot of unsuspected errors. One way to check this is by investigating the OFDM signal at the receiver end. Here the constellation is being verified for errors and inconsistencies.

Deeper analysis is usually focusing on constellation or I–Q analysis, i.e. in-phase and quadrature (I–Q) plane analysis.

Main areas where the constellation analysis (in the I/Q plane) be used are:

- amplitude imbalance
- phase noise
- phase error
- modulation error.

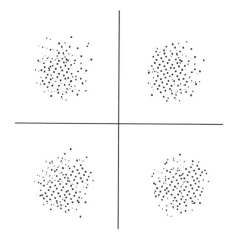

Figure 5.12   Example of the system noise effect interpreted from the constellation figures

System Noise

System noise indicates the interference level of the network (Figure 5.12). A larger spread area means higher interference levels. When the area of one part is spilling into other areas, inter-symbol interference is present.

Phase Noise

It is introduced by local oscillators and timing references within the transmission chain. At the receiver, the phase noise has two different effects. Low-frequency noise gives rise to Common Phase Error (CPE). In this case, all the OFDM carriers are suffering from the same phase error. This error can be removed by the demodulator. High-frequency noise introduces inter-carrier interference (ICI). The noise from one carrier is superimposed on top of neighbouring carriers, and cannot be removed by the demodulator. One type of phase noise is the phase jitter.

Phase Jitter

Jitter in the system is considered as phase noise and is defined by variation in the time domain (Figure 5.13). This means that the two signals are arriving with a phase shift. The carrier regeneration is not possible and the system is producing errors.

Coherent Interference

In coherence interference, the interferer or harmonic components are in phase with the wanted signal (Figure 5.14). This results in ring-shaped constellation figure.

The measurement equipment can be used to identify the performance or problems e.g. in the following elements:

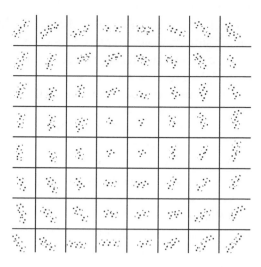

Figure 5.13   Display of the phase jitter

- service protection element: the functionality and status of the element;
- encoders: status and functionality;
- ESG server: the functionality of the element and ESG presence;
- IPE network manager: IP encapsulation status and functionality of the element;
- IP encapsulation and SFN adapter: services status and element functionality;
- transmitter and modulator: RF channel and SFN status, quality of the modulation, status of the element.

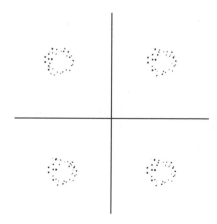

Figure 5.14   Coherent interference causes ring-shaped extension to the decision-making area

The head-end revisions include e.g. the IPE status check, i.e. if the encapsulation is done correctly and making sure there are no packet losses occurring in the encoder side. The respective analyzers can be used in general e.g. for the following: RF layer (received power level or field strength, modulation constellation, guard interval, etc.), SFN cell analysis (channel impulse response), MPEG2 layer (program ID monitoring, MPEG error rate, stream display), DVB-H timing structure analysis, MPE and FEC compliance validation and SI/PSI tables compliance validation as well as for the transmission functionality checking.

For the main transmitters and IPE, the IPE manager monitoring provides central control of all contribution sites in the network in baseband level. The UdiCast Navigator is one example of the tools for the designing of the network and match theoretical coverage calculations with practical coverage. GOLDENEAGLE equipment can be located in main centres to monitor and analysis of DVB-H streamings.

The secondary transmitters can be monitored remotely via equipment to raise alarms in case of problems.

An SNMP tool can be set up for the central monitoring of all equipment in the network. This solution can be used to assure e.g. the commonly agreed service level agreement compliance and the network performance quality in general.

Other measurements can give information e.g. about the TPS Info including DVB-H-specific enhancements and Cell-Id, DVB type, frequency, hierarchy.

The MPEG layer can contain information about the TS Synchronization Loss, TS Synchronization Byte Error, PAT Error, PMT Error, Continuity Count Error, PID Error, Frame Error Ratio (FER), Transport Error, CRC Error and CAT Error.

The MPE layer can contain e.g. MPE Frame Error Ratio (MFER) for all services inside the TS, IP Packet Error Ratio (IP-PER) for all IP packets inside the TS, MPE-FEC check for all services inside the TS, Time Slice/Delta-t Jitter Error measurement for all services.

The IP/UDP layer can consist of cross-check INT with received IPs, Bit-Rate Measurement IP-Port level, RTP Sequence Counter check and SDP analysis and cross-check.

The FLUTE layer can consist of FDT instances (check of announced and received files, instance completion time, cycle counter and times).

The ESG layer can consist of the program and content information, access information, purchase information, SDP files (session description protocol) which consist of decoder information (A/V services), information about additional ESG container, decryption information and information about additional IP sessions (e.g. STKM).

5.6   Other Equipment

The GPS is needed for the synchronization purposes in case SFN mode is used. The synchronization is done by delivering the GPS clock information to the SFN adapter and DVB-H transmitter. The synchronization is especially important in SFN network in order to handle the correct interpretation of the guard interval limits and thus SFN limits.

For the satellite transmission, the related transmitter and receiver are needed. This can be considered as a part of the transport equipment and is basically transparent for the system.

6

Functionality of the System

The functionality of the DVB-H system is in most part based on that of functionality of DVB-T but it has also functionality specific only to DVB-H. The core functionality in the DVB-H system is described in the following sections.

6.1 The Protocol Stack of the End-to-End OMA-BCAST over DVB-H System

The end-to-end OMA-BCAST over DVB-H system provides distribution of audio, video and data over IP which is further delivered over DVB-H. Figure 6.1 depicts the protocol stack of OMA-BCAST over the DVB-H system. The upper layer consists of OSI layers 3–7, where the audio and video streams are carried over UDP/IP by using RTP/RSTP protocols. The data and OMA-BCAST-specific ESG components, in turn are carried over UDP/IP by using FLUTE. The DVB-H bearer limits to the OSI layers 1 and 2, where IP packets are first encapsulated with MPE and associated RS parity data is encapsulated, respectively, by using MPE-FEC. The real-time parameters of the Time Slicing are carried within the MPE section headers. The PSI/SI is a parallel protocol that carries the link layer signaling information. Finally, the aforementioned link layer protocols are split into TS packets and carried as one or up to two transport streams over the DVB-H physical layer. In addition, the modulator adds the TPS in the beginning of each OFDM frame of the transmission.

6.2 Upper Layer Protocols

The upper layer protocols, i.e., the OSI layers 3–7, are specified by OMA-BCAST. The upper layer protocols are related to content delivery, ESG signaling, and

The DVB-H Handbook Jyrki T.J. Penttinen, Erkki Aaltonen, Jani Väre and Petri Jolma
© 2009 John Wiley & Sons, Ltd

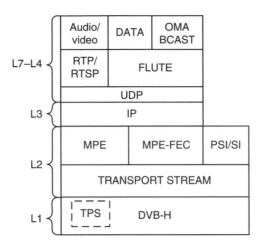

Figure 6.1   Depicts the protocol stack of the OMA-BCAST over DVB-H system

content and service protection. The detailed definition for the upper layer protocols can be found in [OMA08] and [OMA08b].

6.3   MPE in DVB-H

The multiprotocol encapsulation was originally defined for the purposes of DVB data broadcasting in [Dvb06b], as generic section protocol for different DVB systems. The MPE and MPE-FEC sections are based on the structure the DSMCC_section Type "User private" [ISO00], which has already been used within DVB data transmission prior to the definition of DVB-H. The syntax of the MPE was adopted for the DVB-H and the MAC address fields were redefined for the purposes of real-time parameters of Time Slicing and MPE-FEC signaling. Hence, the syntax of the MPE and MPE-FEC sections are mutually identical. The differentiation between the MPE and MPE_FEC sections is done based on the table_id value. In addition to the real time parameters, TS and MPE-FEC-specific signaling is also carried within PSI/SI. The real-time signaling is described in Section 6.4.3 and the TS specific PSI/SI signaling is explained in Section 6.4.4.

6.4   MPE-FEC Frame

MPE-FEC frame is a combination of application data and related parity information of the used FEC code. The MPE-FEC frame is a matrix composed of 255 columns and from 256 up to 1024 rows as shown in Figure 6.2. The maximum size of the MPE-FEC frame is approximately 2 Mbits. It has been split into two parts dedicated to IP datagrams, padding, and parity information of the FEC code, i.e., RS data.

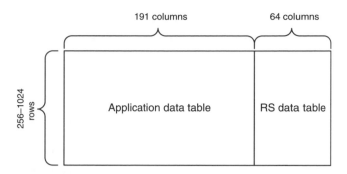

Figure 6.2   The structure of MPE-FEC frame

The IP datagrams and padding are always allocated in the left side of the frame while the RS data is allocated to the right. The left part of the MPE-FEC frame is named as application data table and it consists of 191 rows. The right part of the MPE-FEC frame is named as RS data table and it consists of 64 rows. The data in the application data table is carried within MPE sections of table_id $0 \times 3e$. The parity information in the RS data table is carried within the MPE-FEC sections of table_id 0×78. Each byte within the application data table and RS data table of the MPE-FEC matrix holds information byte, which can be identified with the address that is calculated by multiplying the number of columns with the number of rows. Because the addressing starts from zero, the maximum address value in the application data table would be $191 \times 1024 - 1 = 195583$. The maximum address value of the RS data table, in turn, would be $64 \times 1024 - 1 = 65535$. Hence, for example the byte positioned within the column 150 and in a row 300 would have address value of $150 \times 300 - 1 = 44999$.

6.4.1   Application Data Table

The construction of the application data table is done in a "datagram-by-datagram" basis. Figure 6.3 depicts the principle of the filling procedure of the application data table. The filling procedure is started from the upper left corner and continuing from row to another until all datagrams are allocated for the application data table. The remaining bytes within the application data table are filled with padding, i.e., zero bytes.

6.4.2   RS Data Table

The application data table is used as a basis for calculating the 64 parity bytes allocated into the each row of RS data table. The code used within the calculation is the Reed–Solomon RS(255,191,64). Figures 6.4 and 6.5 illustrate the principle

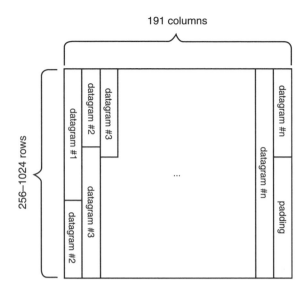

Figure 6.3 The principle of the filling procedure of the application data table

of the RS encoding process. First the RS parity bytes are calculated row-wise by using the bytes in the corresponding row in the application data table. The resulting parity bytes are placed in the RS data table as shown in Figure 6.4. Then the MPE-FEC sections are formed column-wise from the parity bytes as illustrated in Figure 6.5. That is, each RS column is carried in the payload of its own MPE-FEC section.

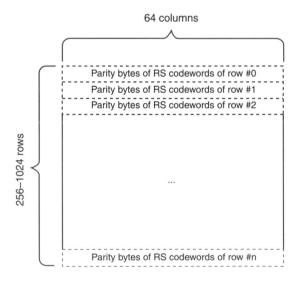

Figure 6.4 The parity bytes of the RS code words within the RS data table

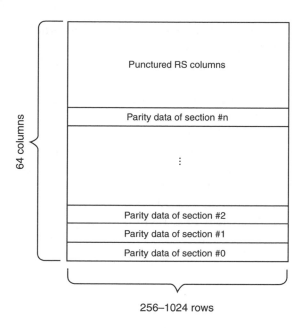

Figure 6.5 The principle of forming the MPE-FEC sections from the RS data table

The code rate of MPE-FEC can be changed from its nominal value, which is approximately 0.75 (191/255), by using code shortening or puncturing. In the code shortening, the code rate can be decreased by adding zero padding columns in the application data table, whereas in the puncturing the code rate can be increased by discarding complete RS columns. There is no signaling needed for the discarded columns, since the receiver is able to detect these when decoding the RS data table. In the code shortening, the number of padded columns is signaled in the header of the MPE-FEC sections (padding_columns).

6.4.3 Real-Time Signaling

The real-time signaling is carried within the section headers of MPE and MPE-FEC sections. The signaling fields carried within the MPE and MPE-FEC sections are identical, but the meaning is different in cases when both, TS and MPE-FEC are used within the same elementary stream and, e.g., when only MPE-FEC is supported for the particular elementary stream. The real-time parameters are as follows.

- *delta_t*: When both, TS and MPE-FEC are applied to the particular elementary stream, the meaning of this field is as follows:
    This field indicates the time to the next Time Slice burst, i.e., delta-t, within an elementary stream. The delta-t resolution is 10 ms. If the delta-t is set to the

value 0×00, it means that no bursts will be transmitted anymore within the associated elementary stream.

If only MPE-FEC is supported within the associated elementary stream, the following applies:

The value of this field increases after each MPE-FEC frame. Hence, this field is used as a counter within the MPE-FEC section headers, to indicate whether there are any frames lost during the transmission.

- *table_boundary*: This field is a 1-bit flag, which is used only when MPE-FEC is supported within an elementary stream. When this field has been set to "1," it indicates that the current section is the last section of a table within the current MPE-FEC Frame, i.e., this filed shall be set to the value "1" for the last MPE section and for the last MPE-FEC section of the particular MPE-FEC frame. If MPE-FEC is not supported by the particular elementary stream, this field shall be reserved to future use and hence not used within that particular elementary stream.
- *frame_boundary*: This field is a 1-bit flag, which indicates that the current section is the last section within the current burst, when this field has value "1." This field can be set to either MPE or MPE-FEC section, if it is the last section of that particular burst.
- *address*: This is a 18-bit field that associates the payload carried within the MPE sections with the RS data carried within the MPE-FEC sections. The address value will start from the value "0." If MPE-FEC is not supported, then this field should not be used and can be ignored.

6.4.4   Time-Slice Specific PSI/SI Signaling

The Time Slice specific signaling in PSI/SI is realized through the Time_Slice_and_FEC_identifier_descriptor. The purpose of this signaling is summarized as follows:

- provides Time_slicing and MPE-FEC labeling for each transport stream and/or elementary stream;
- indicates the frame size used within particular elementary stream or within all elementary stream carried within the transport stream;
- indicates the Max burst duration of particular elementary stream or of all elementary stream carried within the transport stream;
- indicates the Max average rate of particular elementary stream or all elementary stream carried within the transport stream.

The Time_slice_and_FEC_identifier descriptor can be transmitted either within NIT, INT or CAT. When it is carried within CAT or INT, it applies to particular elementary streams and while carried within NIT, it applies to the complete transport streams and hence within all elementary streams carried within those transport streams. The *frame size, Max burst* duration, and *Max average* rate parameters are described in the following:

- *frame_size*: This is a 3-bit parameter having a different meaning when associated with the elementary streams carrying application data and when associated with the elementary streams carrying the RS data, i.e. MPE-FEC labeled for that particular elementary stream.

  If this parameter is associated with the elementary stream carrying the application data, it indicates the maximum number of bits on section payloads, which are allowed within one TS burst of the elementary stream. When this parameter is associated with the elementary stream carrying the RS data, it indicates the exact number of rows on each MPE-FEC frame on the elementary stream. If the elementary stream supports both, i.e., carries application data and RS data, then the maximum burst size and the number of rows are both applied.
- *max_burst_duration*: This 8-bit field indicates the maximum burst duration within the elementary stream.
- *max_average_rate*: This 4-bit field indicates the maximum average bit rate in MPE section payload level over one TS cycle or MPE-FEC cycle.

The Time_slice_and_FEC_identifier descriptor also carries a parameter named *Time_slice_fec_id*. However, it was defined for future usage purposes and is currently set always to the value "0."

6.5   Principle of Time-Sliced Transmission

A time-sliced burst can also be called a channel. Figure 6.6 illustrates a division of channels in time, where there is off-period between each channel. As the lower part of the figure depicts, in the time-sliced transmission the bit rate can also vary between different channels. This allows the allocation of capacity for different types of services.

A channel can be divided into subchannels. It provides further power savings, but the service access delay increases accordingly. As the channel switching time does have a practical maximum time delay in the fluent usage of the services, the definition of the time slicing burst size and the subchannel allocation should be done carefully. It can be estimated that about 2 s of average waiting time in channel switch could be the practical limit the users normally tolerate.

The off-time of the burst is defined by burst size and bitrate. Figure 6.7 depicts an example for the calculation of the TS parameters. The total bitrate assumed for the service is 522 kbps, which includes audio and video bitrate and the bitrate resulted

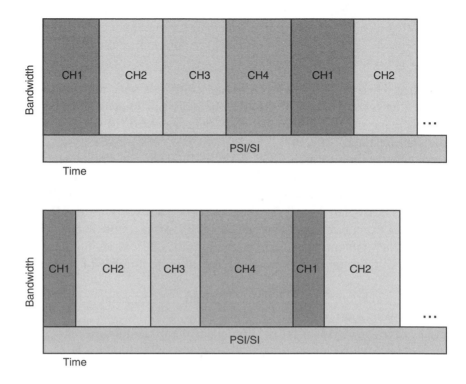

Figure 6.6   The principle of the time-sliced transmission

for the other protocols such as RTP, UDP, IPSEC, and IPv6. The parameters used within the example are as follows:

- burst size $= 1$ Mbit
- video bitrate $= 384$ kbps
- audio bitrate $= 96$ kbps
- other protocol bitrate $= 40$ kbps

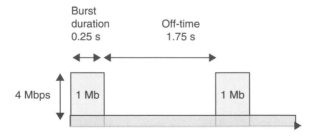

Figure 6.7   An example for the calculation of the TS parameters, where the DVB-H burst duration can be obtained by dividing the burst size with the stream bitrate

- burst size $= 1$ Mb
- burst bandwidth $= 4$ Mbps.

Now, a single burst with 1 Mb window contains about 2 s of TV stream (with $\sim$522 kbps). When the 4 Mbps bandwidth is used, it takes 0.25 s to transmit 1 Mb of data.

   In practice, the channel switching time also depends on the internal terminal processing delays which might be 1–2 s depending on the buffer sizes. The total channel switching time could be thus about 3.5 s in this type of example.

6.6   Program-Specific Information (PSI)/Service Information (SI)

PSI/SI forms the core part of the service discovery signaling within DVB-H and it has been defined in [ETS06b] and [ETS06c]. Together with the ESG information and TPS, it provides means for receiver for signal scan, service discovery and handover functionality. The PSI/SI information is carried in a specific section syntax structure, where the signaling information is carried within several different section types. The PSI/SI is explained in greater detail in Chapter 7.

6.7   Transport Stream

The DVB systems have adopted transport stream as specified within the [ISO00]. The PSI is also part of the [ISO00] and it has been inherited as such.

6.8   Transmission Parameter Signaling (TPS)

The TPS was first defined for the purposes of DVB-T. It was further extended for the purposes of DVB-H. The main purpose for the TPS within the scope of DVB-H is to fasten receiver signal scan and synchronization procedure. The TPS is explained in more detail in Chapter 7.

6.9   The Head-End Functionality

The head-end of the DVB-H system is responsible for the transmission of the service, related service information, and error recovery information. An example of DVB-H head-end is depicted in Figure 6.8, where functionality can be split as follows:

(1) The service system is responsible for providing the services to the system. In addition the service system generates the ESG, which provides description of the services. The service system also provides the application level protocols

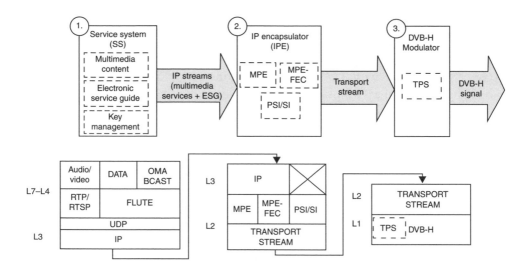

Figure 6.8 The head-end of the DVB-H system

needed for the service discovery and rendering at the top of OSI layer 3. Another
task of the service system is the key management, which is needed for the
service and content protection in case of service encryption is used. The
interface between the service system and IP encapsulator is IP and hence the
service system provides IP streams as input to the IP encapsulator.

(2) The IP encapsulator encapsulates the IP streams into the MPE sections and
calculates the RS parity data for the IP datagrams. The RS parity data is further
encapsulated into the MPE-FEC sections. Another task of the IP encapsulator is
to provide PSI/SI signaling, where the IP addresses are associated with the
location within the transport stream. In addition, network topology information
is provided within the PSI/SI. Finally MPE, MPE-FEC, and PSI/SI sections are
further packetized into transport stream packets. The transport stream is further
provided as output from the IP encapsulator, which is forwarded to the
modulator. If hierarchical transmission is in use, two transport streams are
provided as output, instead of one.

(3) Finally, the modulator generates the DVB-H signal in which up to two transport
streams are carried. Another important task of the modulator is also to provide
the TPS, in which, e.g., the DVB-H indicator and cell identification is provided.

6.10   The Receiver and Terminal Functionality

In the receiving end, two separate entities can be identified. These are called receiver
and terminal. Next, the main functionality of the DVB-H receiver and terminal is
depicted in Figure 6.9.

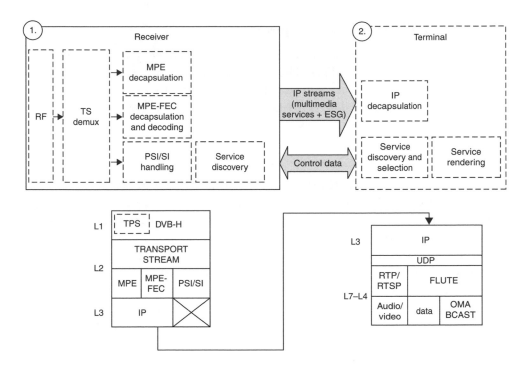

Figure 6.9 The functionality of the DVB-H receiver and terminal

(1) The receiver has RF part, which takes care of the synchronization and demodulation of the DVB-H signal. Next, the TS demux demultiplexes the transport stream and splits it further into the PSI/SI, MPE, and MPE-FEC sections.

In the receiver part, also the MPE and MPE-FEC sections are also decapsulated and the RS data is decoded. The RS data decoded from the MPE-FEC sections is used to repair possible errors within the MPE sections. As a result from the MPE/MPE-FEC decapsulation and MPE-FEC decoding procedures, the IP datagrams are provided as input to the terminal.

Another important functionality of the receiver is the PSI/SI handling in which the PSI/SI subtables are parsed from the received PSI/SI sections. The service discovery information is obtained from the resulted PSI/SI subtables.

(2) The terminal is responsible for the functionality above the OSI layer 3. It processes and decapsulates the received IP datagrams. The IP datagrams carry both, the multimedia services and the service discovery, and access information, such as ESG, carried above the OSI layer 3. The service discovery procedure is handled partially in the terminal and partially in the receiver. In addition, the monitoring of the services and handover functionality needs to be handled within the receiver or in the terminal.

6.11   The DVB-H Network Types

The DVB-H network topology can be either single frequency network (SFN) or multifrequency network (MFN). The transport stream distribution and needed network infrastructure deviate between SFN and MFN deployments. Some example scenarios of SFN and MFN are provided in Sections 6.11.1 and 6.11.2.

6.11.1   Single Frequency Network

The basic form of the DVB-H network is the SFN. If SFN is deployed as network scenario, the same bitstream is distributed throughout the network by using one frequency. The latter means that in case of nonhierarchical modulation, a single transport stream is distributed throughout the network area. If hierarchical modulation is used, one high-priority (HP) and one low-priority (LP) transport streams are be carried throughout the network area. SFN is also requires minimal network infrastructure, since only one IPE is needed to provide transport stream(s) for the entire network. Figure 6.10 illustrates an example of SFN deployment in nonhierarchical DVB-H network. As shown in this figure, only one IPE is needed for

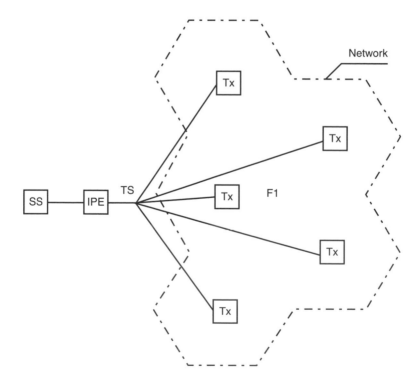

Figure 6.10   An example of the SFN scenario deployment in nonhierarchical DVB-H network

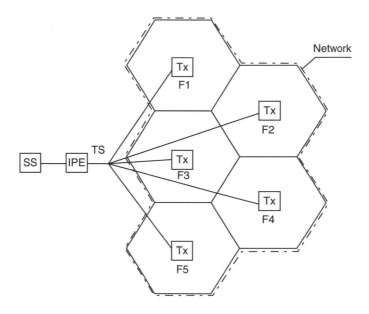

Figure 6.11 An example of the MFN scenario deployment in nonhierarchical DVB-H network, where only one transport stream is distributed throughout the network area

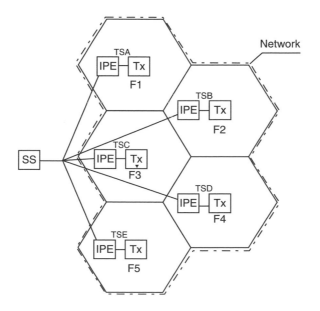

Figure 6.12 An example of the MFN scenario deployment in nonhierarchical DVB-H network, where one transport stream is distributed per cell of the network

generating single transport stream, which is further transmitted throughout the whole SFN area within one frequency.

6.11.2   Multifrequency Network (MFN)

The MFN networks allow distribution of several different transport streams throughout the network area. However, it is also possible to use similar content distribution as in SFN, i.e., transmits one transport stream throughout the network area. The only difference in the latter compared to SFN network scenario is the number of used frequencies. Figure 6.11 illustrates an example of the MFN network, where only one transport stream is distributed throughout the network.

Another possibility for content distribution within MFN scenario deployment is depicted in Figure 6.12, where each frequency of MFN distributes different transport stream. Such scenario allows local content distribution within each cell and also requires local IPE per each cell.

7

Signalling

7.1 Service Discovery Signalling

The service discovery signalling means the signalling that enables the receiver to discover the desired services from the DVB-H signal. The service discovery signalling in the end-to-end DVB-H system has been split into the upper layer and DVB-H bearer. The upper layer provides the description of the service and associates services with IP addresses within the Electronic Service Guide (ESG). The DVB-H bearer, in turn, associates IP addresses of the services with the location within the transport stream carried by the DVB-H signal. In addition, it provides signalling support for the handover. Hence, the IP functions as an interface between the upper layer and the DVB-H bearer. Figure 7.1 illustrates the protocol stack of the IPDC over DVB-H system and the division of the upper layer and DVB-H bearer into different OSI layers.

7.1.1 DVB-H Bearer

The service discovery signalling of the DVB-H bearer consists of PSI/SI and transmission parameter signalling (TPS).

7.1.1.1 Transmission Parameter Signalling

TPS signalling is L1 signalling which provides information of the transmission scheme, cell identification and whether the DVB-H and MPE-FEC are supported within the associated signal.

The DVB-H Handbook   Jyrki T.J. Penttinen, Erkki Aaltonen, Jani Väre and Petri Jolma
© 2009 John Wiley & Sons, Ltd

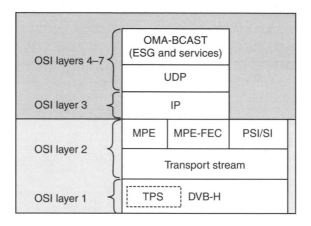

Figure 7.1 The protocol stack and the division of IPDC and DVB-H bearers within an end-to-end DVB-H system

TPS is carried in the beginning of each orthogonal frequency division multiplexing (OFDM) frame. One TPS bit is conveyed for each OFDM symbol and each OFDM frame contains 68 symbols. One superframe, in turn, is composed of four consecutive OFDM frames. The relation of TPS bits, symbols, OFDM frames, and superframe is illustrated in Figure 7.2.

The following information is carried within the TPS signalling:

- Constellation: Indicates the constellation used within the associated signal.
- Guard interval: Indicates the guard interval used within the associated signal.
- Code rate: Indicates the code rate used within the associated signal.
- Transmission mode: Indicates whether the used transmission mode is 2K, 4K, or 8K.

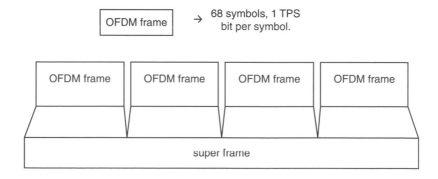

Figure 7.2 The OFDM frame structure

- Super frame number: Indicates the number of the current super-frame.
- Cell identification: Indicates the cell_id of the cell covered by the associated signal.
- DVB-H indicator: Indicates whether the associated signal supports DVB-H.
- MPE-FEC indicator: Indicates whether the associated signal supports MPE-FEC.

7.1.1.2 Program Specific Information (PSI)/Service Information (SI)

PSI/SI signalling is the core part of the signalling in the DVB-H. The main purpose of PSI/SI signalling is to associate IP addresses and the location within the transport streams which are carried within the DVB-H signals. PSI/SI is composed of signalling subtables, which provide information related to the service location within the current and other multiplexes within the network. Each subtable, in turn, may be composed of one or more sections. PSI/SI is transmitted periodically and the repetition of each subtable varies. The sub-tables used within the DVB-H are as follows:

- Network Information Table (NIT)
- Program Association Table (PAT)
- Program Map Table (PMT)
- IP/MAC Notification Table (INT)
- Time and Date Table (TDT).

Network Information Table (NIT)
In DVB-H, the NIT is used for two main purposes. First, it provides information of the current and other network and associates each transport stream with the network, frequency and cell information. One transport stream may be associated with one or more network, frequency and cell. Each frequency and cell, in turn, may carry up to two transport streams. The second purpose of the NIT in the context of DVB-H is to announce and provide location for the available IP platforms within different transport streams of the network. The NIT is transmitted in a fixed packet identifier (PID) value of 0×10.

Program Association Table (PAT)
The purpose of PAT is to associate each service_id with the PID. The PAT is always carried within the fixed PID value of 0×00.

Program Map Table (PMT)
One PMT is allocated for each service_id. The purpose of the PMT is to list the packet identifiers of the elementary streams which are in the scope of the associated service_id. In addition, the PMT associates each elementary stream with the

component_tag. The PID for the PMT is dynamic and can take values from the range of 0×0020 to $0 \times 1FFE$. It is allocated by PAT.

IP/MAC Notification Table (INT)
Each INT subtable is platform specific where IP addresses unique within that platform are announced. The INT is used for associating IP addresses with the location within the transport stream. The location information within the transport stream is provided with five parameters: network_id, transport_stream_id, original_network_id, service_id component_tag. Similarly as in the case of the PMT, the PID value for the INT is dynamic and it is allocated by the PAT.

Time and Date Table (TDT)
The TDT is purposed to provide the up-to-date UTC time to the terminals. The table contains the current network time and date in UTC and MJD.

7.1.2   Upper Layer

The service discovery signalling within the upper layer consists of association of services with the service description and IP address information. The IP address information includes IP source and destination addresses and ports. ESG is purposed for the end user to acquire information about the available services. Figure 7.3 illustrates a general diagram of the upper layer components, where the association between the service, service description, service and content protection and IP address information is provided.

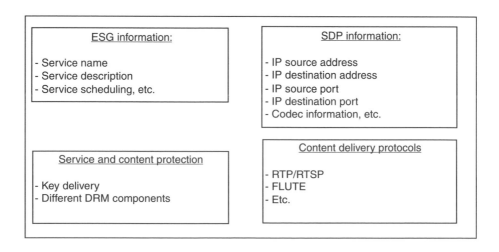

Figure 7.3   The generic diagram of the upper bearer components

7.2 End-to-End Service Discovery

Jani Väre

This section gives a generic example of the service discovery from the initial signal scan until the discovery of the service. The following description focuses on the most important parameters in the service discovery signalling and also reveals some implementation techniques which can be used for the service discovery. Section 7.2.1 provides an overview of the signal scan from the initial phase until INT discovery. Next, in Section 7.2.2 an overview of the service discovery is provided, starting from the discovery of ESG bootstrap and service selection until the discovery of the elementary stream of the selected service.

7.2.1 Signal Scan

Signal scan means scanning of all possible frequencies that are available for DVB-H. The extensity of signal scan may vary depending on the previous information available on the existing DVB-H signals within the area. The receiver may need to scan all signals which could be available or it might need to settle on just a few signals. In any case signal scan is the procedure which needs to be repeated very rarely and hence the importance of the signal scan extensity is minor. The signal scan within DVB-H has been handled various times in the research papers since the first introduction. Figure. 7.4 depicts the general flow for the signal scan procedure followed with the description of each step.

Step 1: The TPS lock can be achieved already a few milliseconds before the receiver has completed the synchronization to the frequency. Hence, the rest of the synchronization procedure can be skipped if the TPS signalling information indicates that the current frequency does not carry any service that is of interest to the receiver.

   The frequency used within this step is determined by the receiver. It may be e.g. start from the first possible frequency within the range and then continue with the next possible until the whole frequency range is exhausted.

Step 2: If the TPS signalling indicates that the frequency carries DVB-H services, then the synchronization is completed and the procedure continues to Step 3. Otherwise the current frequency is discarded and the procedure is started again from Step 1.

Step 3: Once the synchronization is completed, the receiver may start to receive NIT. Typically NIT is received within 10 s but the reception may also be faster or take more time. The NIT reception duration depends on the network configuration and on the robustness within the current location. Once NIT is received, the receiver may store all needed information to the memory, such as location for the desired IP platforms, i.e. INT discovery information. Also, the

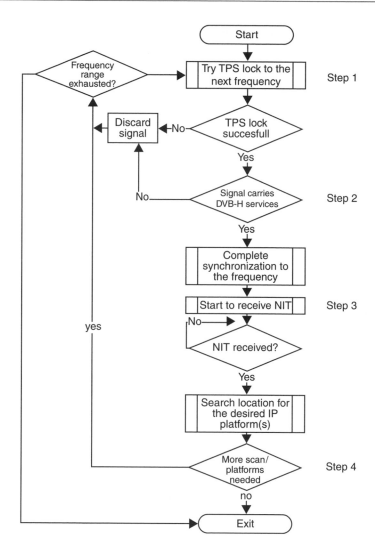

Figure 7.4   General flow of the signal scan

topology information of the network is needed for the handover purposes. The
storing of the NIT information to the memory is advantageous implementation
decision, since the same information can be re-used within the area of the same
network, and if NIT remains the same, without necessarily needing to re-receive
NIT. Figure 7.5 depicts the mapping of INT discovery information within PSI/SI,
where the steps are described as follows:

(1) Linkage descriptor maps one transport stream with one or more IP platforms.
(2) Transport stream is associated with frequency and other OFDM parameters
      through the transport stream loop of NIT.

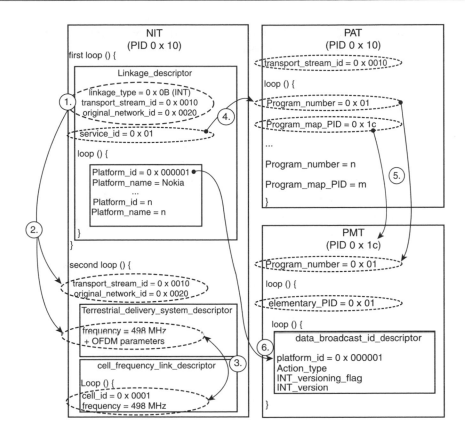

Figure 7.5   The mapping of INT discovery information within PSI/SI

(3) The transport stream loop further maps the transport stream and frequency with the cell information.

(4) IP platform is associated with PMT through service_id and PAT.

(5) Finally, the PID for the INT of the desired IP platform is discovered through PMT PID associated within PAT and platform_id carried within the data_broadcast_id_descriptor and associated with the INT PID within the PMT.

Step 5: Depending on the implementation strategy, the receiver may continue for searching for more IP platforms from the remaining set of frequencies. Hence, the procedure continues from Step 1. Otherwise the procedure can exit.

7.2.2   Service Discovery

The service discovery procedure can be started when at least one DVB-H signal is discovered during the signal scan procedure. The discovery of ESG is a special case in service discovery. The discovery of ESG must always precede the discovery of any other services, since the ESG contains information about the actual availability of multimedia services. The ESG and multimedia discovery procedures are in the most part similar, since in both, PSI/SI is in the main role.

7.2.2.1   ESG Discovery

Similarly as in the case of signal scan, the ESG discovery can be assisted by the previous information, which the receiver has about the sought ESG or the location of the ESG. Both the network and receiver implementation may have impact on this. In general the terminal should have at least the following information about the sought ESG, ESG providerID, ESGprovider name and the platforms where the sought ESG is available. Next, Figure 7.6 depicts the ESG discovery procedure, followed with the description of each step.

Step 1: INT is received and the mapping information is stored to the memory.
Step 2: The transport stream and related service_id associated with the ESG bootstrap IP address is discovered from the INT.
Step 3: The PID of the elementary stream carrying the ESG bootstrap carousel is discovered through the PAT and PMT and filtered by the receiver. Next, the IP address for the desired ESG is discovered from the bootstrap.

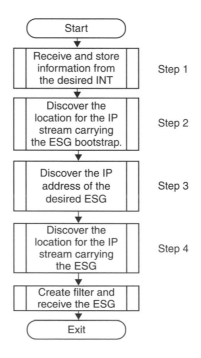

Figure 7.6   The ESG discovery procedure

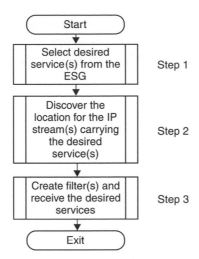

Figure 7.7   The multimedia service discovery procedure

Step 4: The location information for the IP address of the selected ESG is discovered from the INT and the PID for the elementary stream carrying the ESG is sought through the PAT and PMT. Finally the filter is created for the ESG and it is received.

7.2.2.2   Multimedia Service Discovery

The multimedia service discovery starts when the end user selects desired service(s) based on the ESG. Figure 7.7 depicts an example of the discovery of multimedia service where one multimedia service is selected from the ESG. Steps 1–3 are described below.

Step 1:   One or more of the desired services are selected from the ESG.
Step 2:   The location information for the IP addresses of the selected services is discovered from the INT and the PIDs for the elementary stream carrying the services are sought through the PAT and PMT.
Step 3:   Finally filters are created for the services and these are received based on the discovered information.

7.2.3   Service Monitoring and Handover

Once the services are selected and are under consumption, the terminal/receiver needs to take care of the monitoring of the service availability in the possible handover candidates. The service availability in the neighbouring cells is provided within the INT. From the receiver implementation point of view, it needs to receive and decode a new INT each time when handover is performed and in addition to the TS location information within the current cell, it also obtains the TS location of the currently consumed services in the those TSs provided within the neighbouring cells. In some cases, where the PSI/SI transmission is centralized, i.e. the same PSI/SI is provided within each cell, this information may need to be received only once, unless it is changed.

7.3 Interaction Channel

Jyrki T.J. Penttinen

7.3.1 General

DVB-H provides the content delivery only in the downlink direction. It is an adequate solution for the open contents, i.e. free-of-charge type of channels e.g. for the television and radio type of programs.

Even the basic DVB-H definition does not contain the return channel; there are various manners for the interaction between the terminal and the network. In principle, the type of interaction channel does not actually matter technically as long as the relevant information can be delivered from the end user for the DVB-H system. This means that e.g. standard Internet connection via wireless or local area network could be used for the interaction. Nevertheless, the most logical way to inform the system about user actions is via mobile communications systems like 2G or 3G packet data connection as the respective functionality is normally integrated into the same DVB-H terminal. In the case of a stand-alone DVB-H terminal, also the scratch card solution is possible to adopt in order to open the contents.

The interaction can be divided into two main parts: opening the protected source channel context (e.g. selection of the pay TV channels) and sending program-related information in the uplink direction (like televoting and interactive gaming). The interaction can also trigger the billing event depending on the interaction type. In the case of 2G and 3G, the billing as well as the respective authentication and authorization of the user can be utilized as such, which makes the 2G and 3G network even more logical solution for interactions.

The interaction via mobile communications networks can contain TV channel independent services as alerts, communities and value-added services. The TV channel-related services, on the other hand, can contain e.g. televoting and opinion polling, games, quizzes and competitions, and additional information to the television contents in general. The system allows functionalities that give overall added value to the normal contents e.g. by showing normal television advertisement with a temporally appearing internet link, which can be accessed directly with the terminal by clicking the button and initiating point-to-point packet switched call, which launches the related web page giving more information about the topic, with further product ordering links, etc. The system provides a sufficiently good platform for almost endless solutions that can be used in relation to the television/radio program and related additional services in an interactive way.

Currently, there are two main evolution paths for the wireless access method as for the interactions, OMA-based and Conditional Access (CA) type of solutions. OMA (Open Mobile Alliance) means using open standards whilst CA is of proprietary type. CA has been traditionally the solution for the fixed TV users whilst OMA-based access follows the basic philosophy of the mobile telecommunications solutions.

Although the DVB-H standards do not limit the physical layer of the interaction channel type, the 2G and 3G mobile services are thus the most logical way of handling the DVB-H uplink communication. The packet switched GPRS (General Packet Radio Service) that is defined to both GSM and UMTS provides the straightforward solution as the DVB-H handset contains normally also DVB-H module either as an integrated or as a separate HW/SW module.

The DVB-H functions without additional elements for delivering the contents that can be observed by the end user without any interactions whatsoever. This basic mode is suitable for the free-to-air channels, but if the service provider would like to charge for the specific selection of closed channels, there should be a separate interaction channel integrated to the DVB-H core network.

7.3.2   Interaction via the Mobile Network

If the return channel is done via the mobile network, the authentication, authorisation and accounting (AAA) of the DVB-H user can be done using the already existing infrastructure and other related services of the network. It is thus straightforward to use the SIM-based procedures of the mobile network.

In the planned DVB-H coverage area, the issue of the return channel arises from the fact that the radio part of the mobile and DVB-H networks are different due to the differences in the radio parameters and characteristics of the radio channels (Figure 7.8). The common sites for the mobile system and DVB-H can be used, but the coverage areas differ from each other due to the different power levels, antenna heights, frequency, modulation schemes and other parameter settings.

It can be assumed that packet switched connections of GSM and 3G are used in the major part of the cases for the DVB-H return channel. In order to function correctly, the return channel must thus be dimensioned accordingly as the coverage and capacity is considered.

The quality of the GSM/UMTS interaction data channel depends on the level of the overlapping coverage areas of DVB-H and GSM/UMTS networks. In addition, the general functioning of the data channel affects also the end-to-end performance

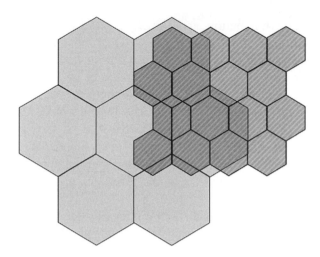

Figure 7.8 The three different cases of the coverage area of the mobile network return channel within or outside of the DVB-H coverage

of the interactions done for the DVB-H content petitions. In other words, if the GPRS network quality is weak (due to e.g. capacity or coverage problems), the situation does not get better in case of using it as an interaction channel for the DVB-H.

The overlapping of DVB-H and mobile network coverage can be divided into three major categories:

- Areas where both DVB-H and GPRS function with sufficient quality.
- Areas where good-quality DVB-H service is found but with only poor or non-existing GPRS service.
- Areas where no DVB-H service is found although GPRS service exists.

The first option is the ideal one for the fully functional interactive DVB-H service set. This means that the contents can be managed at each moment via GPRS e.g. in order to open the closed contents or for entering the point-to-point data service using the URL appearing in the DVB-H hand held.

The second option provides with the sufficient means to follow the DVB-H contents. If the closed contents have been already opened (i.e. the protection key change procedure has taken place) there is no problem in using the respective channels, or open channels. Nevertheless, e.g. the televoting is not possible due to the lack of the uplink data channel.

The third option is the least functional for the broadcast data service as the DVB-H coverage does not exist in respective area. Nevertheless, there is still possibility of using the streaming services via mobile network services (e.g. MBMS), or via point-to-point data services with fully functional interaction services.

Concentrating on the first option, i.e. the fully overlapping DVB-H and GPRS coverage areas, the quality of the GPRS is playing the main role in end-to-end performance when making the interactions.

For the DRM and simple interaction usage, even a basic GPRS service is sufficient. The capacity of a typical GPRS network is sufficient for providing relatively low bit rates for multiple GPRS users. Each GSM TDMA time slot can be multiplexed for up to eight simultaneous users. The possible increment of the interference levels of the GSM/GPRS network means in reality a small reduction of the GSM/GPRS coverage area during the most occupied moment. According to the theoretical studies, the effect could be in the order of approximately 2 dB/Pen99/. This means basically that in the normally dimensioned GSM network (with some non-BCCH frequencies in downlink), there can be a small fluctuation, i.e. cell breathing effect, noted depending on the cell load. In normal, not too tightly dimensioned overlapping GSM/GPRS networks this is not a practical problem, though.

The UMTS radio transmission is based on the WCDMA, or in the later phase of Long Term Evolution (LTE), on OFDM. Even if the evolution of the mobile networks brings each time higher data rates, the data speed is not limiting factor in DVB-H type of interactions as the message size is minimal. The added value of the higher data rates can be seen though at the moment when the user switches from the DVB-H broadcast reception to a specific topic using point-to-point connection.

7.4   Service and Content Protection

Jyrki T.J. Penttinen and Kyösti Koivisto

The delivery of the DVB-H contents in free-to-air mode is quite straightforward as the digital rights items are not considered. Nevertheless, when the user wants to access a protected channel, the contents must be delivered together with rights access elements with the possibility to the end user to access the content and charge the usage of it in a controlled way.

7.4.1   Securing the Content

There are different possibilities of running the DVB-H business as described in the third chapter of the book. In the mobile operator-driven model, Mobile Operators aggregate their own mobile TV service portfolio by making deals with TV channels and content owners, whereas in the broadcaster-driven model, broadcasters acquire capacity from the network operator and offer the services directly to the consumer. In the practical set-ups, this model is the most likely one if mobile TV services are offered free-to-air, but the model also allows pay TV services.

In addition to the correct charging of the content usage, it is also important to have a trust model between the companies running the DVB-H business. This means that the content is delivered, charged and used in an agreed way. As there are various business models that can be applied to the DVB-H operations, including dynamic way of offering the contents to the customers e.g. with the possibility to extend the rights during certain special promotion periods, it is important that corresponding methods for the content protection can be found in the market. The market is also developing in such way that it is possible to deliver the contents for the end users via several telecommunication networks, which means that the digital rights should be independent of the technology used for the content delivery.

The traditional solution for the pay-TV is to use conditional access (CA) for the service protection and the access to the television contents. The typical way of applying CA is to construct it inside of the terminal with corresponding software. The actual handling of the rights can be made by using an access card with the corresponding chip. This is a suitable solution for the traditional rights management for a television set or a set-top box which is big enough to accommodate an additional smart card reader. On some systems the recorded content is recorded using the original encryption (together with the corresponding access keys) and thus the CA is able to control the further use of the recorded contents.

7.4.2 OMA Model

The mobile network environment has additional aspects due to the mobility itself as well as for the protection of mobile terminal specific services, ringtones and logos, and differs thus from the fixed television reception. OMA DRM 1.0 (Open Mobile Alliance Digital Right Management) is the industry standard for protecting a wide range of mobile-related items, including video clips.

OMA DRM has already been in use for the protection of the mobile content. In order to adapt the method for the evolution of the mobile communications, OMA has further developed the original DRM specification to OMA DRM version 2.0. The clearest benefit of the OMA solution is that it is non-proprietary, open standard which does not cause the fragmentation of the DRM market. This leads to the economic scalability as the open standard can be implemented to a wide range of terminals allowing for interoperability.

OMA BCAST used OMA DRM 2.0 specification to develop a system for protecting mobile video. This specification is known as the OMA BCAST DRM Profile. An almost identical system was defined by DVB and that is called 18Crypt. In addition, OMA BCAST developed a system that uses the SIM card features available in mobile phones. This system is known as the OMA BCAST Smart Card Profile.

In summary, OMA BCAST specification for the service purchase and protection, together with 18Crypt, is a flexible, network-agnostic and future proof solution for the content security for the mobile TV environment.

DVB specified the so-called Open Security framework, which has proprietary elements allowing current CA providers to reuse their proprietary key management systems and broaden their business to DVB-H. DVB also designed an open, fully standardized system, 18Crypt, which uses only standardized components for key and secrets management. OMA choose to use 18Crypt as a basis for the DRM profile system. Thus 18Crypt and OMA DRM Profile are very close to each other when the technical functionality is considered.

The OMA DRM Profile is a SPP standard that has been optimized especially for the broadcast media, and it can be used for any IP-based contents during the transmission of the streams as well as for the content protection after the transmission, i.e. OMA DRM 2.0 can also be used with Service Purchase and Protection (SPP) methods to manage the media consumption in mobile terminals.

The 18Crypt has been developed by major industry representatives of broadcasters, mobile network operators, component manufacturers, security companies and terminal vendors, and it has been approved as a standard by ETSI and IEC. The 'family-tree' of the DVB-H DRM solutions is depicted in Figure 7.9.

The OMA BCAST specified profiles and 18Crypt provide all the relevant purchase methods that are used in pay-TV solutions The subscription-based payment supports variable subscription time if the related restriction is activated. The program-based payment limits the access to a certain selected program that the user wants to consume. In event-based payment, the access is limited to a certain event, meaning e.g. a bundle of services related to a sports event. Pay per view offers the possibility to purchase based on time tokens. Preview provides means for accessing the contents during a limited period. Figure 7.10 illustrates the functions of the OMA BCAST DRM profile and 18Crypt.

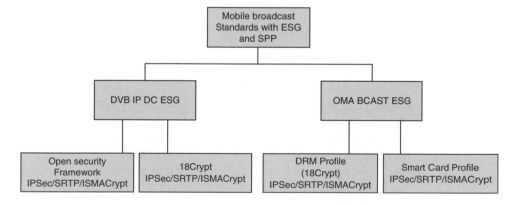

Figure 7.9   The 'family-tree' of the DVB-H DRM solutions

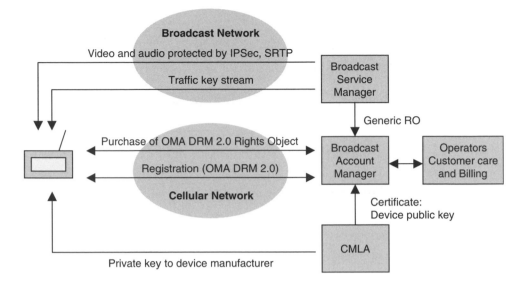

Figure 7.10   The functions of the OMA BCAST DRM profile and 18Crypt

IPSec is one of the encryption methods defined for the Smart Card Profile, DRM Profile and 18Crypt. IPSec can be used for IP-based content providing a secure transmission of the content. The dominant encryption method has turned out to be Ismacryp. All OMA BCAST specified profiles as well as 18Crypt support Ismacryp as an encryption method. The DRM profile and 18Crypt use OMA DRM 2.0 for so-called Key Messages as illustrated in Figure 7.10.

As OMA specified solutions are fully specified standards, they do not leave variations in the implementation and provide thus a solid base for the interoperation of products, service systems and terminals. The obvious benefit of the interoperability is the possibility to adopt various business models, including both DRM and Smart Card profile. The picture in Figure 7.10 illustrates this. DVB has, though, noticed that shortcoming of IPDC ESG and recently specified an adaptation layer also for Smart Card profile.

As a family of specifications, OMA BCAST specified standards allow operators to start with DRM profile and later upgrade to Smart Card profile, if a need is foreseen. The ESG does not need to be changed as it provides elements to support all OMA BCAST specified systems, allowing a flexible path to system enhancements.

In addition to the local and national content delivery, also roaming is important in mobile TV environment, providing possibilities for the additional business. The default expectation of the customer is to be able to use Clear-to-Air (CTA), Free-to-Air (FTA) and encrypted services. The standardized DVB-H solution offers the same air interface between terminal and visited network as is in the customer home network, enabling the roaming function as long as mobile network operators

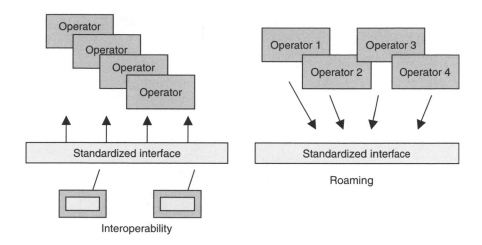

Figure 7.11   The generic diagram of the standardization interface and operator relation in the case of interoperability and roaming

have an agreement that allows service roaming. The roaming in a broadcast network means, of course, access to broadcasted local content, not content from roamer's home network. Yet this feature is of utmost importance for a customer when travelling or especially when participating big events (e.g. sports ones) which have good local TV coverage. Figure 7.11 illustrates the generic diagram of the standardization interface and operator relation in the case of interoperability and roaming.

OMA DRM 2.0 specified Rights Objects (RO) are used to deliver keys and entitlements. The security of the DRM Profile and 18Crypt is based on widely adapted AES-128 and RSA algorithms. The key lengths have been selected to provide optimal terminal performance and security.

One of the important aspects of DVB-H and its security-related functionality is the possibility of using both IPv6 and IPv4 networks and enable protection to all contents transmitted over the broadcast network. IPSec guarantees that any content delivered on IP could be encrypted. In addition to IPSec, also SRTP (Secure Real-time Transport Protocol) and ISMACryp may be used. While IPSec provides merely protection for service delivery, SRTP can also be used to protect stored content. IPSec is a basic element of any IP stack implementation of an IP receiver, and SRTP is a mandatory element of 3G devices. Therefore they do not generate a need for any additional SW components.

There are several ways by which terminal system software can be upgraded in the field. As an example, the Nokia N92 and N77 can use an OMA standardized FOTA system to enable system software upgrades. The upgrade can be limited to sections of software or to cover the full software thus offering flexible system upgrades.

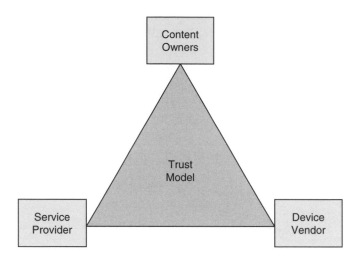

Figure 7.12   The CMLA trust model

The terminal SW update mechanism checks that the origin and the source of the new SW are approved and of verified source.

All OMA DRM-based systems use the CMLA (Content Management License Administration) trust model (see Figure 7.5). CMLA is a non-profit limited liability corporation that has been founded by Intel, Matsushita, Nokia and Samsung to implement a trust model and PKI for systems based on the OMA DRM 2.0 technology. Recently CMLA has widened its scope to cover also mobile broadcast.

CMLA as a trust model defines a widely accepted legal framework to ensure content security in download and broadcast services.

The CMLA role can be seen as the insurance provider for the SPP system. CMLA is the third party that will neutrally analyze all problems. If misconduct is identified, CMLA has the legal means to demand for implementation improvements and in case all other efforts fail, use their legal power for liability payments and in the utmost case, terminal revocation. This is clearly an improvement to the situation where one company would decide where the fault is and if liabilities are due.

CMLA provides the necessary trust model, i.e. the legal framework defining the level of robustness for all implementations, and appropriate penalties for vendors not being as robust as defined (Figure 7.12). It also issues keys and certificates for devices and rights issuers and runs on-line parts of the PKI system.

8

DVB-H Head-End

8.1 Overview

Mobile TV head-end is the centralized point where the DVB-H signal is generated. High-level architecture of basic DVB-H head-end is shown in Figure 8.1. The figure shows an architecture that is based on real deployments rather than reference architectures used in different standards.

Content is coming in from the left side and the generated DVB-H signal is going out from the right side. At the bottom there is a connection to billing and charging system in order to charge DVB-H subscribers.

8.2 A/V Content Encoding

Available A/V content is typically broadcast TV content. Existing TV content is formatted for large screens and it usually has different encoding than DVB-H uses. Therefore existing content cannot be used as it is. Instead, it needs to be encoded for DVB-H use.

Encoders typically support different A/V input formats. Most of the encoders support analog formats for audio and video but those are not usually used for commercial deployments. However analog inputs are useful for testing because analog interfaces are widely supported by cheap consumer products.

In the commercial environment digital A/V input is commonly used. The highest possible quality is achieved by raw unpacked digital audio and video; typical interface is SDI that can carry one video channel and a few audio channels. Good quality is also achieved by MPEG2 encoded digital audio and video; typical interface

The DVB-H Handbook   Jyrki T.J. Penttinen, Erkki Aaltonen, Jani Väre and Petri Jolma
© 2009 John Wiley & Sons, Ltd

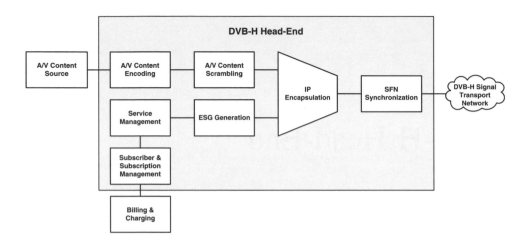

Figure 8.1 DVB-H head-end architecture

is ASI that can carry multiple audio and video channels. MPEG2 encoded content can be alternatively transported to encoder over IP which is becoming more and more common transport. Figure 8.2 shows one example of encoding input and output.

H.264 codec has been specified to be used for DVB-H video encoding and AAC/HE-AAC/HE-AAC-v2 has been specified to be used for DVB-H audio encoding. Both audio and video codecs have lots of parameters. Each TV/radio channel can have its own set of parameters independently of each other. Usually parameter values are fine tuned based on required quality versus available capacity.

In UDP/IP level encoder RTP output for one A/V channel is actually 4 RTP/RTCP streams. Audio and video content is carried by RTP packets. A RTCP stream provides out-of-band statistics and control information for a RTP stream. RTCP is not useful in DVB-H broadcasting. However, standard streaming client requires RTCP flows; otherwise client will stop receiving the streaming. Therefore, RTCP flows have to be sent to the receiving terminal. Typically in IP level RTCP flows consume less than 0.5% of the total content streaming capacity, so RTCP provides very little capacity overhead. Figure 8.3 shows encoder RTP/RTCP output streams for one TV channel.

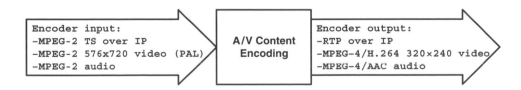

Figure 8.2 Encoding input and output example

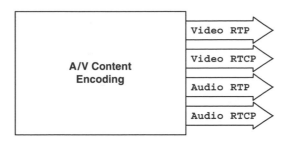

Figure 8.3   Encoder RPT/RTCP output streams

IETF has specified how MPEG4 content is transported over RTP [RFC 3640].
Figure 8.4 shows how content encapsulation has been specified; IP version 4 is used
in the example.

RTP packet is transported over UDP/IP, which has fixed header size. RTP payload
includes an AU header section, auxiliary section and AU data section. Header and
auxiliary sections are optional; typically with DVB-H AU header section is used but
an auxiliary section is not used.

The AU header section includes mandatory fixed size AU header length field
which defines total header length in bits excluding the length field itself. Optional
padding bits are included to make sure that the total length of the header fits the octet
boundary. The AU data section includes one or more AUs; each AU has a respective
header in the AU header section.

Finally, the AU header consists of different fields. Order of fields has been
specified, and the existence and length of each field is signaled out-of-band to
terminal. In practice session SDP configuration is delivered inside ESG and that also
includes the AU header field existence and length information. Receiving terminal
needs this information for header decoding.

As an example, encoder session information for audio contains the following
parameters:

```
SizeLength=13; IndexLength=3; IndexDeltaLength=3
```

The length of one AU header is 16 bits and no padding bits are needed. Encoding
of the captured audio packet is shown in Figure 8.5.

In the above example there is only one MPEG4 audio AU per IP packet, auxiliary
section is not used and padding bits are not needed. In the AU header only AU size
and AU index/AU index delta fields are used.

IP packet contains 243 bytes of MPEG4 audio payload and 44 bytes of header
information. If this IP packet presents the average packet of the flow, it can be
assumed that on top of the MPEG4 payload there is 18% of header overhead for this
particular audio RTP packet flow. If the respective audio encoding speed is 64 kbps,

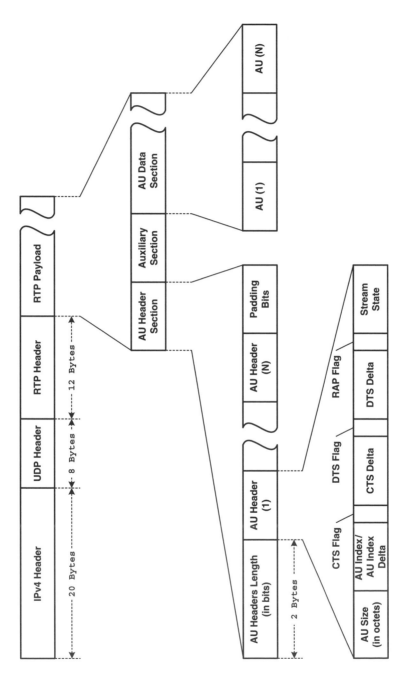

Figure 8.4 MPEG-4 content encapsulation

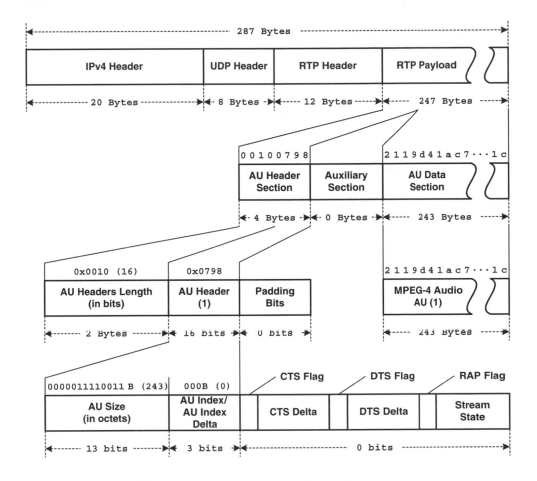

Figure 8.5 Encoding example of one audio packet

calculated IP throughput in encoder output for audio RTP packet flow would be 76 kbps. In order to calculate total IP throughput in encoder output for one particular channel, video RTP flow and both RTCP flows need to be taken into account too.

8.3 A/V Content Scrambling

Because of the broadcasting nature, all the receivers will receive exactly the same content. To control the access to the distributed content, service and content protection needs to be used. This can be implemented by encrypting the video and audio content and using the selected key delivery mechanism to make sure that decryption keys are delivered to end users who have subscribed to the service.

OMA BCAST service and content protection specification [BCAST10-ServContProt] list the following encryption technologies to be used with DVB-H:

IPsec [RFC2406], SRTP [RFC3711], and ISMACryp [ISMACRYP11]. ISMACryp is used in the following example.

Scrambler receives encoder output streams as an input, scrambles audio and video RTP streams and additionally creates respective key stream. Figure 8.6 shows scrambler output streams for one TV channel.

ISMACryp specifies that AES [AES] in counter mode shall be used as the encryption cipher. Both AES blocksize and key length are 128 bits. The amount of data remains the same before and after encryption.

ISMACryp header is defined by the following pseudocode:

```
if (ISMACrypSelectiveEncryption) {
   bit(1) AU_is_encrypted;
   bit(7) Reserved;
}
else AU_is_encrypted = 1;
if (auNum==0) // First AU in packet?
{
   unsigned int (ISMACrypIVLength * 8) initial_IV;
   unsigned int (ISMACrypKeyIndicatorLength * 8)
   key_indicator;
}
else
{
   int (ISMACrypDeltaIVLength * 8) delta_IV;
   if (ISMACrypKeyIndicatorPerAU)
      unsigned int (ISMACrypKeyIndicatorLength * 8)
      key_indicator;
}
```

As an example, scrambler session information for audio contains the following parameters:

```
ISMACrypSelectiveEncryption=0; ISMACrypIVLength=4;
ISMACrypDeltaIVLength=0; ISMACrypKeyIndicatorLength=2;
ISMACrypKeyIndicatorPerAU=1
```

Selective encryption is not used, IV length is 4 bytes and key indicator length is 2 bytes. As a result, the length of one ISMACryp header is 48 bits. Encoding of the captured encoded audio packet is shown in Figure 8.7. Input for the scrambled audio packet below is exactly the same as that was shown in Section 8.2.

Scrambler inserts ISMACryp header into the RTP header section before AU header. In this example ISMACryp header contains only IV and key indicator

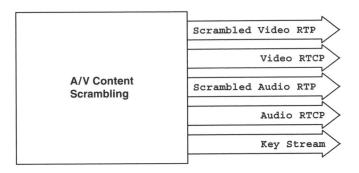

Figure 8.6   Scrambler output streams

fields as defined earlier in the session information. As a result, IP packet contains 243 bytes of encrypted MPEG4 audio payload and 50 bytes of header information. On top of the MPEG4 payload there is 21% of header overhead for this particular scrambled audio RTP packet flow. If the audio encoding speed is 64 kbps, the calculated IP throughput in scrambler output for audio RTP packet flow would be 77 kbps.

8.4   Electronic Service Guide (ESG) Generation

ESG is the mandatory component for the DVB-H terminal for proper content reception. ESG contains information of how IP streams are mapped into the MPEG2 transport stream. ESG may also contain other service-related information like schedule, program, and service purchase information. In practice, ESG is set XML files multicasted to receiving terminals over the DVB-H broadcast. ESG is mandatory element of service discovery and it is not encrypted.

ESG generator takes related service information as an input, creates ESG XML files and multicast those to IP encapsulator. File-based multicasting uses FLUTE/ALC protocol over UDP. File sending is done continuously and it can be illustrated as a file carousel (see Figure 8.8).

When a file is updated, it replaces respective file in the carousel and it will be sent in the next round

8.5   IP Encapsulation

IP encapsulation bundles multiple IP multicast streams onto a single elementary DVB MPEG-2 stream. Generated constant bit rate transport stream is ready to be broadcasted into the air. Protocol stack is shown in Figure 8.9.

IP packets are encapsulated by using MPE frames. MPE framing adds 16 byte header overhead for each IP packet. Figure 8.10 shows the MPE header structure.

DVB SI/PSI signaling information is also included into the transport stream. SI/PSI data is capsulated directly into the MPEG-2 transport stream. SI/PSI information

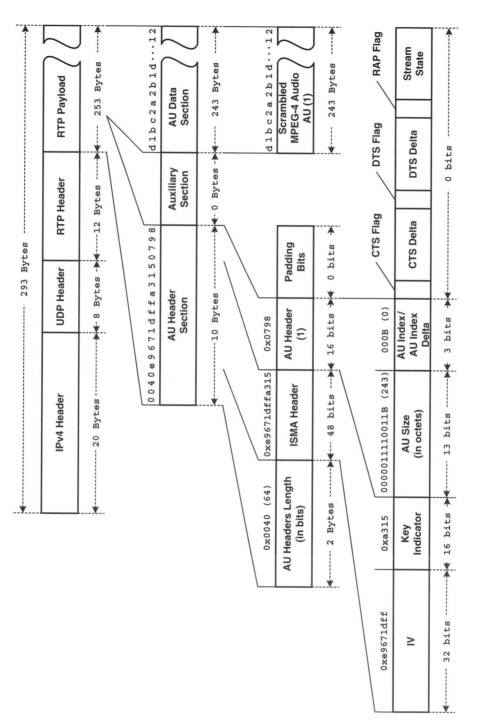

Figure 8.7 Encoding example of one scrambled audio packet

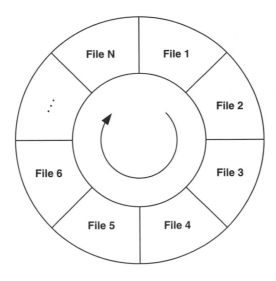

Figure 8.8   File carousel

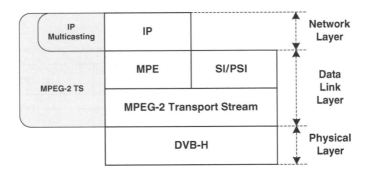

Figure 8.9   DVB-H protocol stack

data model is set of tables containing signaling information. Broadcasting of the tables is continuously repeated.

MPEG-2 transport stream is formed by fixed size MPEG-2 transport packets. Transport packet header size is 4 bytes and it can carry 184 bytes of payload. The MPEG-2 transport packet structure is shown in Figure 8.11.

8.6   Service Management

ESG generation, encoding, scrambling and IP encapsulation require service information. Service management can be implemented on each functional element

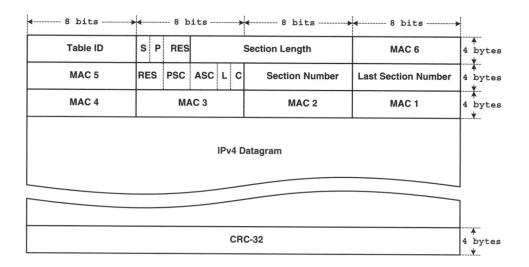

Figure 8.10   MPE header structure

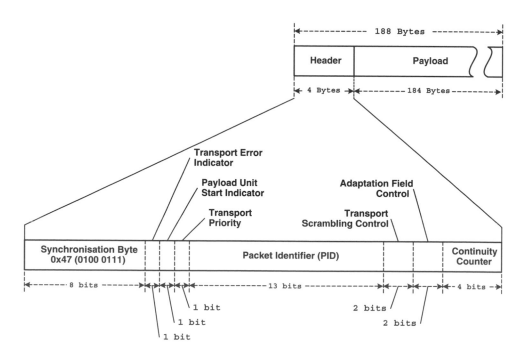

Figure 8.11   MPEG-2 transport packet structure

or on one centralized service management element. Centralized solution is easier to manage because DVB-H service information needs to be configured only once in one place. In practice, service management can be combined where some elements are using centralized information repository and some elements have own dedicated information storage.

Example of information required by ESG generation:
- programs
- program descriptions
- schedules
- mobile operator codes (MCC and MNC)
- purchase information for each mobile operator.

Example of information required by encoding:
- codecs
- bitrates
- frame rates
- sampling frequencies.

Example of information required by scrambling:
- ECMG information
- key lengths
- salt keys.

Example of information required by IP encapsulation:
- DVB parameters
- SI/PSI signaling parameters.

8.7   Subscriber and Subscription Management

Subscriber and subscription management is needed in order to provide commercial chargeable services to DVB-H users. Subscriber and subscription management provides subscription information toward charging and billing system and it may also provide viewing rights to the DVB-H receiver.

As an example, subscription can be pay-per-view type of one time purchase, one or more channels for certain time period or continuous subscription of one or more channels.

8.8   SFN Synchronization

Unused frequencies are very rare resources today. To save frequencies, the SFN broadcasting system has been developed where multiple transmitters are using the same frequency.

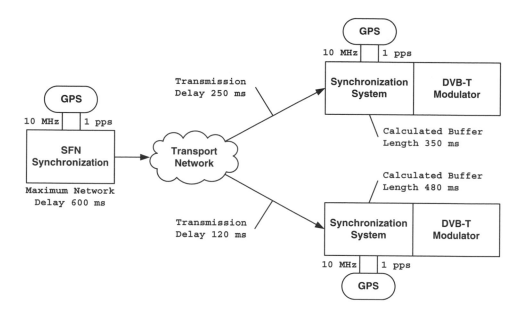

Figure 8.12   Dynamic MPEG-2 transport stream buffer calculation for SFN synchronization

SFN cell consists of multiple synchronous transmitters, each transmitting the same bit at the same time. Because of highly accurate synchronization, the terminal can receive the same signal from multiple transmitters at the same time without getting disturbed. Distance difference from the terminal to different transmitters causes different reception delay of the signal from different transmitters. This delay difference cannot exceed certain value, otherwise reception will be disturbed. The maximum diameter of one SFN cell depends on the broadcasting parameters and it can be up to 65 km.

The broadcasting signal SFN network has to be synchronized in the DVB-H head-end. This is done by periodically inserting synchrony frame into the MPEG-2 transport stream. This is called MIP frame and it includes maximum network delay, timestamp, and some other broadcasting parameters.

Synchronization system on each transmitter calculates transmission delay by comparing current time with timestamp from MIP frame received from MPEG-2 transport stream. Required transport stream buffering is then calculated by decrementing transmission delay from maximum network delay. Buffer adjustment can be done dynamically since transmission delay may vary over time.

As a result, all SFN transmitters are transmitting same signal at the same time. Example of automated dynamic MPEG-2 transport stream buffer calculation for SFN synchronization is shown in Figure 8.12.

9

DVB-H Radio Network

9.1 OFDM

The special behaviour of the OFDM (Orthogonal Frequency Division Multiplex) that DVB-H uses provides important possibilities but also has limitations which are essential to know in the detailed radio network planning and optimization.

OFDM is a spread spectrum technique that has been in use since the 1960s, first in military applications, and later, in commercial telecommunication networks. US Bell labs was the initiator in this field. The modulation system that would be sufficiently robust against radio interface-related phenomena was studied later in CCETT (Centre Commun d'Etudes en Télédiffusion et Télécommunication), and the outcome of the work was Coded OFDM, i.e. COFDM. The technique is especially suitable for environments where the response of the channel varies depending on the frequency. When the receiver interprets the signals (carriers), their level of neither amplitude nor angle is not identical when the frequency varies. The COFDM spreads the information over a wide frequency range, using closely spaced frequency sub-bands. Furthermore, these data streams are coded in transmission side. The COFDM means basically these two functionalities. The radio propagation fading, i.e. the signal level drop due to multipath propagation, occurs randomly and only on few narrow sub-bands at a time.

Coded OFDMA can be described by defining its components:

- FDM spreads the data on sub-bands, and the data is correct on all the sufficiently good sub-bands.
- Coding adds protection codes, which provides a means for the recovering of the data on the bad quality sub-bands.

The DVB-H Handbook   Jyrki T.J. Penttinen, Erkki Aaltonen, Jani Väre and Petri Jolma
© 2009 John Wiley & Sons, Ltd

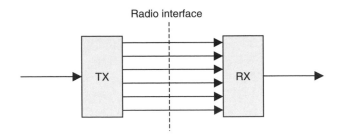

Figure 9.1 OFDM technology delivers the high speed bit stream via several sub-carriers each containing lower speed bit streams. The transmission is parallel, and sub-carriers are separated by different frequencies

OFDM is a multicarrier system where the data is transmitted in parallel sub-channels by using several carriers (see Figure 9.1). Each carrier is digitally modulated, and the modulation scheme can be selected from QPSK, 16-QAM and 64-QAM in the case of the DVB-H.

In coded OFDMA, it is necessary to minimize the interference level between the sub-band signals. It can be generalized that the orthogonality requires that the sub-carrier spacing equals the inverse of the symbol duration. The carrier frequencies are thus chosen so that the spacing between two adjacent carriers is the inverse of symbol duration:

$$f_k = f_0 + \frac{k}{T_u}, \quad k = 0, \ldots, n.$$

As the spectra are overlapping, the carriers must be orthogonal to each other in order to provide the separation, as can be seen in Figure 9.2.

The OFDM transmission is parallel and sub-carriers are separated by different frequencies as shown in Figure 9.3. The information is distributed by interleaving it to the multiple sub-carriers, first having added the error protection, resulting in Coded Orthogonal Frequency Division Multiplex (COFDMA).

The orthogonality of OFDM is observed over a symbol period. In the orthogonal case, the sum of all the sub-carrier signals that are integrated over one symbol period results a zero. Also any two sub-carriers are orthogonal, meaning that when integrating any two sinusoidal functions with frequencies being integer multiple of the sampling rate results to zero. The orthogonality is maintained if both signals are synchronized.

The orthogonality in the frequency domain has the following aspects. The power spectrum of a modulated sinusoidal is a sync function which decays in the frequency domain. The sync function has nulls which fall into the centre frequency of the

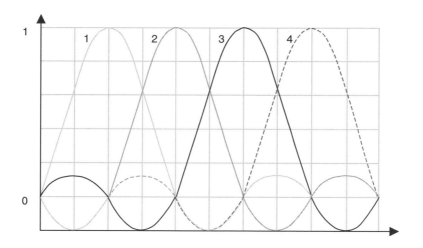

Figure 9.2 The principle of OFDM carrier amplitudes and spacing

adjacent sub-carriers. The side lobes of all adjacent sub-carriers cancel each other. Orthogonality is maintained if the sub-carriers have no frequency errors.

When multipath propagated components occur in the radio interface, the OFDM has the following characters. The components change in the amplitude of the signal, because the multipath adds or subtracts the level in the time domain. This causes

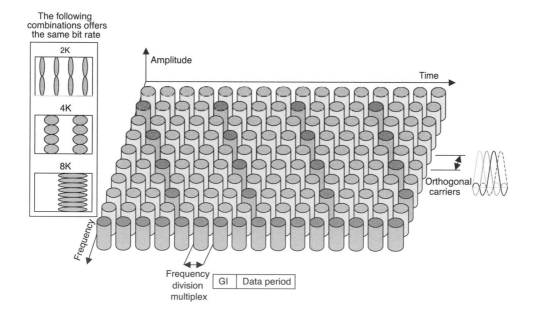

Figure 9.3 OFDM transmission principle

phase distortion. In addition, long multipath delays start to cause inter-symbol interference (ISI).

In OFDM, the guard interval (GI) is a solution for the phase distortion. The ideal GI should be long enough to capture all the multipath components before symbol reception starts. After the guard time, the phase is almost stable. OFDM can also be used with the Cyclic Prefix (CP), which is a technique to copy the remaining symbol shape for a duration of guard time and attach in front of the symbol. This part of the symbol (partial replica) provides estimation of the phase during the multipath vulnerable period.

9.2   DVB-H Radio Frequency

The typical DVB-H frequency is in the UHF band of 470–700 MHz. DVB-H can be built to VHF III (174–230 MHz), UHF IV/V (470–862 MHz), L-Band and TDD IMT bands (1900–1920 MHz and 2010–2025 MHz). The functional frequency can be optimized by taking into account the size of integrated antennas (lower the frequency, bigger antenna is needed) and on the other hand, the larger coverage of lower frequencies. The antenna gains with lower frequencies would be inadequate.

It should be noted that the UHF channels up to number 49 (i.e. less than about 700 MHz) are the most usable if GSM 850 or GSM 900 system is built in the same terminal. This is due to the interferences caused by the GSM transmitter in higher frequencies. Otherwise there is no limitation for the upper terminal frequency. As this has been identified as a limitation, it would be recommendable to take care that at least one DVB-H multiplex can be built nationwide with channel number less than 49 in UHF.

A bandwidth of a single UHF channel of DVB-H is 5–8 MHz. As an example, if the bandwidth is 8 MHz, the channel with 602 MHz centre frequency means a band of 598–606 MHz. Figure 9.4 shows possible VHF and UHF frequency bandwidths of DVB-H. In practice, only some channels are reserved for DVB-H; mainly in the UHF band.

The channel raster (frequency separation, i.e. the bandwidth of the channel) of DVB-H is country dependent, and can be selected between 5, 6, 7 and 8 MHz.

The formulae given in Tables 9.1 and 9.2 show the method of calculating the DVB-H frequencies. In this format, the term f offset $= 1$ Hz in the SFN case, and not bigger than 500 Hz in MFN.

Figure 9.4   The frequency band of DVB-H in VHF and UHF

Table 9.1   The channel raster formulae for VHF III band

Channel raster (MHz)	F_c (centre frequency)	N (VHF channel number)
6	$177.0\,\text{MHz} + (N - 7) \times 6\,\text{MHz} + f\,\textit{offset}$	7, 8, 9, 10, 11, 12, 13
7	$177.5\,\text{MHz} + (N - 5) \times 7\,\text{MHz} + f\,\textit{offset}$	5, 6, 7, 8, 9, 10, 11, 12
8	$178.0\,\text{MHz} + (N - 6) \times 8\,\text{MHz} + f\,\textit{offset}$	6, 7, 8, 9, 10, 11, 12

Note: Please note that in some countries, an offset may be used for VHF. The preferred offset is $\pm n \times 1/6\,\text{MHz}$, where $n = 1, 2, 3, \ldots$

Table 9.2   The channel raster formulae for UHF IV and V bands

Channel raster (MHz)	F_c (centre frequency)	N (UHF channel number)
6	$473.0\,\text{MHz} + (N - 14) \times 6\,\text{MHz} + f\,\textit{offset}$	14, \ldots, 83
7	$529.5\,\text{MHz} + (N - 28) \times 7\,\text{MHz} + f\,\textit{offset}$	28, \ldots, 67
8	$474.0\,\text{MHz} + (N - 21) \times 8\,\text{MHz} + f\,\textit{offset}$	21, \ldots, 69

9.3   Modulation

Modulation schemes that can be used in DVB-H are QPSK, 16-QAM and 64-QAM. QPSK (Quadraphase-Shift Keying) has 2 bits/symbol. 16-QAM (16-state Quadrature Amplitude Modulation) has 4 bits/symbol and 64-QAM (64-state Quadrature Amplitude Modulation) has 6 bits/symbol.

In general, 16-QAM and QPSK are the most functional modes in DVB-H environment, 64-QAM being very sensitive for the interferences. 16-QAM needs about 6 db stronger signals compared to QPSK in order to function correctly. 64-QAM needs around 10 dB stronger signal compared to QPSK.

Even if 64-QAM is challenging in the normal operation of DVB-H, it can be used for offering the hierarchical radio transmission.

If the capacity requirement is not high, i.e. the maximum of about 10–15 audio/video channels per frequency band, QPSK offers the largest coverage areas.

9.4   FFT Mode

In DVB-H, there can be about 2000, 4000 or 8000 carriers. This is expressed as a FFT mode of 2K, 4K or 8K, respectively. Each sub-carrier takes care of the delivery of only a fraction of the bit stream. As an example, 10 Mbps stream with 2K FFT mode means that each sub-carrier actually transports $10\,\text{Mb/s}/2000 = 5\,\text{kb/s}$.

Each FFT mode has pros and cons. The 2K supports fast moving terminals but results small SFN areas. Fast moving terminals require more resistant FFT mode than pedestrian or slowly moving terminals. From the three possible values of the DVB-H FFT mode, the 2K carrier mode is the most Doppler-shift tolerant. It might be a suitable solution e.g. in separate SFN cells along motorways and train rails.

Table 9.3 Example of the *C/N* requirements of FFT modes of DVB-H (the *C/N* requirements for the observed cases)

Case	Modulation	Code rate	Bit rate	*C/N* (dB), Rayleigh channel
1	QPSK	1/2	4.98	5.4
2	QPSK	2/3	6.64	8.4
3	16-QAM	$\frac{1}{2}$	9.95	11.2
4	16-QAM	2/3	13.27	14.2
5	64-QAM	$\frac{1}{2}$	14.93	16.0
6	64-QAM	2/3	19.91	19.3

It should be noted that the small SFN leads to small transmitter coverage areas, which might result in an expensive network if done in large areas.

On the other hand, the 8K FFT mode supports slow moving terminals and large SFNs. Large SFN makes also bigger transmitter sites possible, which results in a more economical solution. These two modes are supported also in DVB-T.

The 4K is a new addition to DVB-H, and is meant to be a compromise between 2K and 8K FFT modes. 4K thus provides sufficiently fast terminal speeds in relatively large SFN areas, making it an economically attractive solution. As the FFT dictates the maximum terminal speed and theoretical SFN size, the final selection of the FFT mode is one of the important optimization tasks in the radio network planning. Chapter 11 shows an analysis, how the theoretical SFN size can be extended in a controlled way by balancing the interference level caused by extending the SFN limits, and on the other hand, taking into account the increased level of SFN gain.

Tables 9.3–9.6 summarise the carrier per noise requirement for 2K, 4K and 8K when $\frac{1}{4}$ guard interval is used in a typical urban 6 km/h radio channel type.

The type of FFT size is subject to speed limitations: 2K and 4K FFT sizes are used for very high speeds, and 8K FFT size is used for relatively slow speeds.

In most cases 8K is enough as it supports speeds up to 150 km/h when MPE-FEC is used. MPE-FEC lowers the useful bitrates due to error correction overhead, but on the other hand, without MPE-FEC mobility requirement can be more challenging to fulfil.

Table 9.4 The *C/N* and Doppler limits for the 2K case

Case	*C/N* min (dB)	Fd max (Hz)	At *C/N* Fd min + 3 dB: Fd (Hz)	At *C/N* Fd min + 3 dB: 500 MHz (km/h)
1	13.0	201	169	365
2	16.0	167	135	291
3	18.5	142	114	246
4	21.5	113	96	207
5	23.5	90	75	162
6	27.0	52	39	84

Table 9.5   The C/N and Doppler limits for the 4K case

Case	C/N min	Fd max	At C/N Fd min + 3 dB: Fd	At C/N Fd min + 3 dB: 500 MHz
1	13.0	133	112	242
2	16.0	111	90	194
3	18.5	96	77	166
4	21.5	74	63	136
5	23.5	60	50	108
6	27.0	36	27	58

Table 9.6   The C/N and Doppler limits for the 8K case

Case	C/N min	Fd max	At C/N Fd min + 3 dB: Fd	At C/N Fd min + 3 dB: 500 MHz
1	13.0	65	55	119
2	16.0	55	45	97
3	18.5	50	40	86
4	21.5	35	30	65
5	23.5	30	25	54
6	27.0	20	15	32

9.5   Guard Interval

The guard interval (GI) of the DVB-H is the pause between symbols on the OFDM sub-carrier. The GI length is relative to symbol duration: 1/4, 1/8, 1/16, 1/32.

At the same time when the GI protects the signal due to the overlapping, it also limits the maximum useful diameter of a SFN cell, i.e. the maxim area of the coverage formed by the set of all radio transmitter coverage areas that belong to the same SFN area. GI thus defines the maximum distance between the extreme transmitters within one SFN area, i.e. DVB-H cell.

In theory, the delay from one edge to another extreme edge in a cell should be less than the distance which the signal propagates during the guard interval.

More carriers mean that the symbol rate on each sub-carrier is lower. As an example, if 2K and 8K systems have the same total bitrates, the sub-carriers on 8K have lower data rate (because there are four times bigger amount of sub-carriers to carry the data). This means logically that the symbols and guard intervals are longer in 8K than 2K. The 8K OFDM mode thus allows larger cells (longer GI means longer maximum distance of transmitters in one SFN). Figure 9.5 shows the principle of the GI in the time scale, i.e. for a single site, the transmission consists of the useful data and space reserved for the GI.

Figure 9.5 Principle of the symbols on an OFDM sub-carrier. As long as the symbols do not overlap (tanks to guard interval) they can be received correctly

It can be generalized that the shortest GI values are not practical due to the much reduced SFN sizes. The values of 1/4 and 1/8 are feasible for DVB-H. Guard Interval is signalled to the terminal in TPS (Transmitter Parameter Signalling).

9.6 Error Correction

The errors in DVB-H reception occur normally only on some carriers at a time. In other words, the interferences of the radio interface are bursty in nature and they are time and frequency selective. For minimizing the effects of interferences in the radio interface, the DVB-H contains inner and outer coding.

The inner coding is referred to as Viterbi coding, which is actually a convolutional coding method that also DVB-T utilizes. It is meant to minimize the effect of the impulse interferences, increasing the tolerance against the bursty interferences. In DVB-H, it is possible to select the related code rate (CR) between the values of 1/2, 2/3, 3/4, 4/5, 5/6 and 7/8. The value indicates directly the fraction of the useful payload bits. In practice, it is recommended to select the value from the ones that are most protected, i.e. 1/2, 2/3 or 3/4. The practical field tests show that the least protected modes are not optimal in the moving environment. On the other hand, the mode 1/2 reserves half of the radio interface capacity for the error coding, so the final decision of the coding rate is again one of the many optimization tasks of the DVB-H operator.

The outer coding is referred to as the Reed–Solomon (RS) error correction method. Its length is 16 bytes. The RS error coding is added for each PDU. The MPEG2 Transport Stream PDU of DVB-H is 188 bytes long, and the PDU means Protocol Data Unit (= packet in a DVB transport stream). In addition, the Inner Interleaving protects against selective fading across OFDM sub-carriers, distributing the bits over several carriers whilst the outer Interleaving minimizes the impact of bursty errors as the protected section is spread out in time. Code rate (CR) refers to the amount of inner Viterbi coding

The error correction of DVB-H is carried out with the RS coding. It is based on the polynomial correction method. The polynomial is encoded for the transmission over the air interfacc. If thc data is corrupted during the transmission, the receiving end can calculate the expected values of the data within the certain limits that depend on the settings. MPE-FEC is an additional error coding method in DVB-H. It has been defined as an optional feature in the terminal side. If the terminals do not support the MPE-FEC correction, it simply rejects the related frames and acts as a DVB-T in error coding.

The differently sized MPE-FEC frames are needed for the optimal planning of the radio network. The value depends e.g. on the number of services, capacity of each service, the on and off periods of the time slicing, etc. It should be noted that the maximum buffer size for the frame is approx. 2 Mbit ($8 \times 255 \times 1024$).

The MPE-FEC that DVB-H uses is based on the RS code. The DVB-H Implementation Guidelines [Dvb06] explains the method, with the possibility of using puncturing. On the other hand, the decoding is left open and the receiver solutions might thus differ from each other. There are various possibilities to decode the signal as indicated in [Him06] and [Dvb09].

The IP datagrams of DVB-H are encapsulated into the MPE-FEC frame. The time slicing burst size affects the MPE-FEC in such way that the amount of MPE-FEC rows can be set to 256, 512, 768 or 1024. The datagrams are encapsulated column-wise to the frame, and the encoding is done row-wise with RS variant RS(255,191), indicating that the total size of the frame is 255 columns, which consists of 191 columns of application data (IP datagrams) and 64 columns of RS data (parity bytes sections and possibly punctured RS fields). Figure 9.6 shows the principle of the frame. The final MPE-FEC rate depends on the filling of the data and RS columns, and can be selected from the values of 1/2, 2/3, 3/4, 5/6 or 1/1 (i.e. when no MPE-FEC is used).

The IP datagram consists of IP header (20 bytes) and IP payload (maximum of 1480 bytes). The IP datagram then consists of MPE header (12 bytes) and CRC-32 check field (4 bytes). On the other hand, the FEC sections (columns) consists of FEC header (12 bytes) and CRC-32 tail (4 bytes). These MPE and FEC packets (with respective MPE or FEC headers) are then fragmented to a transport stream

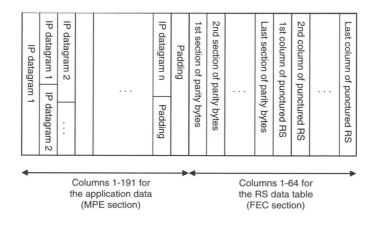

Figure 9.6 DVB-H frame structure

(TS) in such way that TS header (4–5 bytes; 5 if the TS packet contains the first byte of a section) is sent first, then MPE payload (183 bytes) followed by FEC TS header (4 bytes) and FEC payload (184 bytes).

The frame error quality criterion of 5% has been used in DVB-H radio network quality and limit estimations. The 5% criteria can be studied when MPE-FEC is not used (frame error rate, FER), or with MPE-FEC involved (MFER, frame error rate after MPE-FEC). The 5% criterion was originally selected in subjective way in the early phase of DVB-H evaluations, as it provides with practical means of estimating the quality. It is based on the intuitive criteria of maximum of 1 erroneous frame with 1 s of length during 20 s of measuring period, which is still considered as sufficiently good reception quality.

The MPE-FEC performance can be studied in function of different radio parameters, in different radio channel types and terminal speeds. The 5% frame rate before and after MPE-FEC functionality is a common observation point in order to seek for the performance limits. The difference of these curves gives the indication of the MPE-FEC gain as shown in Chapter 11.

In order to carry the DVB-H IP datagrams of the MPEG-2 Transport Stream (TS), the Multi Protocol Encapsulator takes care of the encapsulation of each IP datagram into single MPE section. Elementary Stream (ES) takes care of the transporting of these MPE sections. Elementary Stream is thus a stream of packets belonging to the MPEG-2 Transport Stream and with a respective program identifier (PID). The MPE section consists of 12 byte header, 4 byte CRC-32 (Cyclic Redundancy Check), as well as a tail and payload length.

The main idea of MPE-FEC is to protect the IP datagrams of the time sliced burst with an additional link layer Reed–Solomon parity data. The RS data is encapsulated into the same MPE-FEC sections of the burst with the actual data. The RS part of the burst belongs to the same elementary stream (MPE-FEC section), but they have different table identifications. The benefit of this solution is that the receiver can distinguish between these sections, and if the terminal does not have the capability to use the DVB-H-specific MPE-FEC, it can anyway decode the bursts although with lower quality when it experiences difficult radio conditions.

The part of the MPE-FEC frame that includes the IP datagrams is called the application data table (ADT). The ADT has a total of 191 columns. In case the IP datagrams does not fill completely the ADT field, the remaining part is padded with zeros. The division between the ADT and RS table is shown in Figure 9.7.

In the DVB-H system, the number of the RS rows can be selected from the values of 256, 512, 768 and 1024. The amount of the rows is indicated in the signalling via the Service Information (SI). RS data has a total of 64 columns.

For each row, the 191 IP datagram bytes are used for calculating the 64 parity bytes of RS rows. Also in this case, if the row is not filled completely, padding is

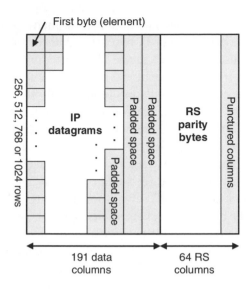

First byte (element)

256, 512, 768 or 1024 rows

IP datagrams

Padded space

Padded space

Padded space

Padded space

RS parity bytes

Punctured columns

191 data columns   64 RS columns

Figure 9.7 The MPE-FEC frame consists of application data table for IP datagrams and Reed—Solomon data table for RS parity bytes

used. The result is a relative deep interleaving as the application data is distributed for the whole bust.

The RS data of DVB-H is sent in encoded blocks with the total number of m-bit symbols in the encoded block $n = 2m - 1$. With 8-bit symbols the amount of symbols per block is $n = 2 8 - 1 = 255$. This is thus the total size of the DVB-H frame.

The actual user data inside of the frame is defined as a parameter with a value k, which is the number of data symbols per block. Normal value of k is 223 and the parity symbol is 32 (with 8 bits per symbol). The universal format of presenting these values is $(n, k) = (255, 223)$. In this case, the code is capable of correcting up to 16 symbol errors per block.

RS can correct the errors depending on the redundancy of the block. For the erroneous symbols whose location is not known in advance, the RS code is capable of correcting up to $(n - k)/2$ symbols that contains errors. This means that RS can correct half as many errors as the amount of redundancy symbols is added in the block.

If the location of the errors is known (i.e. in the case of erasures), then RS can correct twice as many erasures as errors. If N_{err} is the number of errors and N_{ers} the number of erasures in the block, the combination of error correction capability is according to the formula $2N_{err} + N_{ers} < n$.

The characteristic of RS error correction is thus well suited to the environment with high probability of errors occurring in bursts, like in the DVB-H radio interface. This is because it does not matter how many bits in the symbol are erroneous – if multiple errors occur in byte, it is considered as a single error.

It is also possible to use other block sizes. The shortening can be done by padding the remaining (empty) part of the block (bytes). These padded bytes are not transmitted, but the receiving end fills in automatically the empty space.

FEC (Forward Error Correction) is widely used in telecommunication systems in order to control the errors. In this method, the transmitting party adds redundant data to the message. On the other side, the receiving end can detect and correct the errors accordingly without the need to acknowledge the data correction. The method is thus suitable for especially uni-directional broadcast networks.

FEC consists of block coding and convolutional coding. RS is an example of the block coding, where the blocks or packets of bits (symbols) are of fixed size whereas convolutional coding is based on bit or symbol lengths.

In practice, the block and convolutional codes are combined in concatenated coding schemes; the convolutional coding has the major role whereas block code like RS cleans the errors after the convolutional coding has taken place.

In mobile communications, the convolutional codes are mostly decoded with the Viterbi algorithm. It is an error-correction scheme especially suitable for noisy digital communication links. It is used e.g. in GSM, dial-up modems, satellite communications and in 802.11 LANs. It is also included in DVB-H radio transmission. The idea of Viterbi is to find the most probable sequence, Viterbi path in the information flow, i.e. in the sequence of observed events.

The MPE-FEC has been designed taking into account the backward compatibility. The DVB-T demodulation procedure contains error correction so both DVB-T and DVB-H utilize it for the basic coding with Viterbi and Reed–Solomon decoding, whereas DVB-H can also use a combination of Viterbi, Reed–Solomon and additional MPE-FEC to improve the C/N and Doppler performance. The detection of the presence of MPE-FEC is done based on the single demodulated TS packet, as its header contains the error flag. MPE-FEC adds the performance in moving environment as it uses so-called virtual interleaving over several basic FEC sections.

10

Radio Network Dimensioning

When the core network dimensioning is done correctly, i.e., sufficient capacity is reserved between the encoders and radio sites without major delivery delays and the fluctuations (jitter) of data streams, the final performance of the DVB-H will be a result of proper radio network planning. Although the core network dimensioning is important with adequate data stream settings, the radio planning defines the final coverage, capacity, and quality that the end users will experience.

In order to dimension the radio network in optimal way, the coverage and capacity estimations are essential for the initial phase as well as for the longer term evolution of the network. Although the radio parameters could be changed relatively easily in the operational network, the respective effects on the performance must be analyzed thoroughly in these situations. As a rule of thumb, if the customers have already got used to certain quality level and quality, the reduction of either one of these is considered as an annoying factor.

The planning of the DVB-H radio network includes the initial and target quality of the network. These steps are related to the nominal and detailed network planning processes as discussed in the following sections. In addition, it is recommended to understand the effects of the possible interaction channel, which can be done, e.g., via GPRS network. Finally, in the optimization of the DVB-H network, the field measurements are essential.

10.1 Radio Network Planning Process

The DVB-H radio interface includes a set of parameters. Their values dictate the final coverage and capacity with the required quality level (quality of service, QoS).

The DVB-H Handbook   Jyrki T.J. Penttinen, Erkki Aaltonen, Jani Väre and Petri Jolma
© 2009 John Wiley & Sons, Ltd

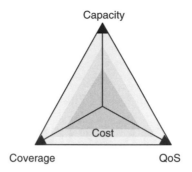

Figure 10.1   The most important interdependencies of the DVB-H radio network

DVB-H parameters normally have interdependencies; therefore the final balancing of radio parameter values requires knowledge about the complete functionality of the network.

In addition to the technical performance impact, the variation of parameter values normally has a cost impact. As an example, the same coverage area of a single cell can be achieved by using a low-power transmitter and high antenna mast, compared to the high-power transmitter and low antenna mast. Obviously, the cost structure of these solutions differs from each other, which should be taken into account in the final dimensioning and deep-level optimization of the radio network.

Figure 10.1 presents the high-level principle of the interdependencies of the DVB-H radio network.

The cost-efficient balance of the above-mentioned items is one of the main tasks in the detailed DVB-H radio network planning (Figure 10.2). As an example, if the coverage is identified as the priority goal with the same QoS and price, its increment automatically means that the capacity of the system must be reduced. A concrete example would be to select QPSK modulation instead of 16-QAM, which could give roughly 6 dB more path loss margin to the link budget, but at the same time the QPSK offers only about half of the capacity compared to 16-QAM.

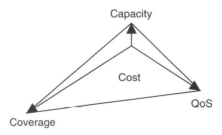

Figure 10.2   The extended coverage means lower capacity with the same quality and cost targets

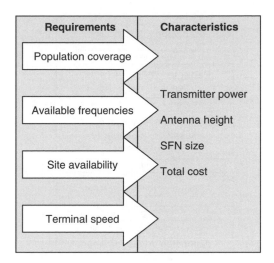

Figure 10.3   The network dimensioning of DVB-H has various inputs

Before starting the planning process, it is essential to collect the relevant data in order to comply with the requirements. As an example, Figure 10.3 shows some requirement categories that impact the network characteristics.

Figure 10.4 gives an example of the requirements collection and related main parameter decision to be used as inputs for the nominal and detailed planning processes.

10.1.1   Service Parameters

It is important to understand in advance, what is the number of services (e.g., audio/video streams) to be provided with the respective quality. Best effort estimation is sufficient in the nominal plan, and more detailed information is needed in the detailed planning process, preferably identifying the most probable extension plans for the capacity also in longer run. The decision of the provided channel number is one of the main items in the beginning of the planning, although DVB-H provides sufficient flexibility for the definition of bit rates per radio channel, i.e., it is possible to define different bit rates for different channels independent of the single channel size.

In the coverage planning of the DVB-H radio network, the requirement for the outdoor and indoor coverage level should be known in early phase. The respective QoS level depends on the area location probability in the single radio cell area (which can also be mapped to the location probability in the radio cell edge). In the nominal phase of the planning, a relatively rough estimation is sufficient.

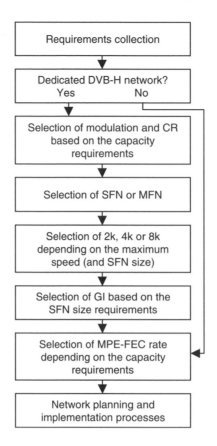

Figure 10.4   An example of how the planning process could be presented in an organized way

10.1.2   Technical Parameters

In the nominal plan, the size of the radio network can be estimated by assuming an average, single value for the antenna height of sites. With this information, the first coverage plan can be produced based on the general radio propagation models. In a sufficiently good nominal plan, different area types should be taken into account accordingly by dividing the analysis to, e.g., dense-urban, urban, suburban, and rural area types. The average attenuation of each area type can be obtained via suitable radio path propagation models. When the estimation for single-cell coverage area per area type is known, it is possible to obtain a rough estimation of the number of needed sites in total service area that is to be covered. Even if this is only a first hand estimate with theoretical propagation models and simplified and averaged assumptions for the cell radius, it also gives estimate for the total cost of the radio network as for the needed equipment and installation work.

In order to estimate the cell radius in different environments, in addition to the propagation loss, the required carrier per noise level is obtained by combining the main parameter values, which are modulation, FFT mode, guard interval, MPE-FEC rate, code rate, transmitter power level, antenna gain, site losses, and possible other effects as SFN gain.

The modulation type should be selected based on the capacity and coverage requirements. As a rule of thumb, QPSK provides the largest coverage areas but with the lowest capacity. 64-QAM requires highest carrier level reducing considerably the coverage area, but on the other hand it provides the highest capacity – 16-QAM is a good compromise for these modes giving the possibility to build the cost-effective long-term network.

The maximum terminal speed depends on the FFT mode. It also affects the maximum SFN size together with GI parameter. In practice, all FFT modes with MPE-FEC provide sufficient performance for the Doppler; therefore in practice, 8K mode is also probably good enough in typical situations in moving environment excluding perhaps high-speed trains when the signal (resultant vector) is arriving straight ahead or behind the train. For the high-speed environments, 2K or 4K is more suitable as can be observed from Chapter 9, but it reduces the SFN size of the network accordingly.

MPE-FEC enhances the error recovery performance depending on the area type (radio channel type). It can be assumed that MPE-FEC gives the best performance in sufficiently fast-moving cases, and there are more multipath radio components involved in order to "wake up" the functionality. It also helps to correct the errors in impulse noise type of environment. On the other hand, the MPE-FEC rate consumes the useful data capacity accordingly with direct proportion.

Channel coding rate is essential in DVB-H in order to recover the errors in moving environment. Correctly balanced values of CR and MPE-FEC rate give the optimal performance in accordance with the achieved user data rate. Even there are studies available, e.g., in Chapter 11 of this book, the final values should be investigated in case-by-case basis as the local environment might have special effects on the performance.

Transmitter power level and the site antenna height dictate the coverage area of the single cell. This should be balanced according to the network costs. This means that the same coverage area can be achieved with low-power and highly located antennas compared to high transmitter power levels and low antenna heights. The respective cost depends on each case as the resulting number of sites with all the related equipment varies. The highest power levels and highest antennas might not be the optimal solution because of the potential interference risk in SFN case, as well as the increased CAPEX due to more complicated equipment and higher energy consumption with more complicated maintenance work, respectively. The antenna height does also have practical values and limitations depending on the area type.

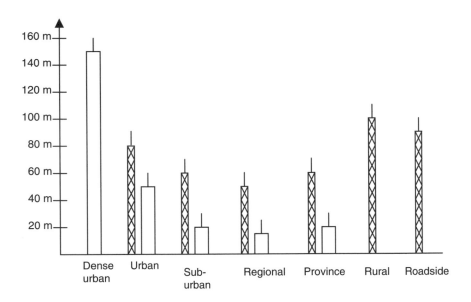

Figure 10.5 Typical DVB-H antenna installations with the respective estimation of the antenna height. The rooftop antenna values depend on the maximum building heights per area

For example, in the city centers, the roof tops might be a good solution with limited power levels due to the radiation restrictions. In the large city environment, the highest buildings can offer an attractive solution for the coverage in the densest city center areas with relatively strong down-tilted antennas, as the signal would attenuate considerably in the center due to the buildings and other obstacles if the site is far away from the centre.

Figure 10.5 shows a basic principle about the average antenna heights in different environments. The values naturally depend heavily on the more specific factors like city size. In this case, the dense urban means the biggest metropolitan centers with skyscraper type buildings.

As an outcome, the nominal plan gives a rough idea about the expected coverage area size with assumed transmitter site distances from each others (the density of sites per planned area), maximum SFN size (in case of the SFN network), total capacity of the area as well as the estimated costs of the solution.

10.1.3   The Balancing of the Radio Planning Items

In the detailed planning phase of DVB-H radio network, there are several items that should be considered while balancing between the network quality, coverage, and costs. Figure 10.6 identifies the most significant aspects. It is important to conduct an economical analysis for the technical options in order to identify the most cost-efficient solutions. This might mean several iteration rounds by varying the relevant

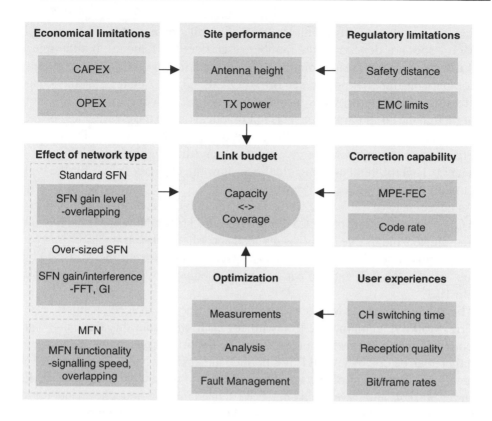

Figure 10.6   The main components in the DVB-H radio network planning

technical parameter set values and the revision of the price of each solution. Nevertheless, even the task is somewhat time consuming, it is worth carrying out in early phase of the planning process as it have a major impact on the total network costs.

10.2   Capacity

In the initial phase of the DVB-H network planning, the offered capacity of the system is estimated. The total capacity in the defined DVB-H band – defined as 5, 6, 7, or 8 MHz – does have effect on the size of the coverage area as well. The dimensioning process is iterative, with the aim to find a balance between the capacity, coverage, and the cost of the network.

The capacity can be varied by tuning the modulation, GI, CR, and channel bandwidth. As an example, the parameter set of QPSK, GI $\frac{1}{4}$, CR $\frac{1}{2}$, and channel bandwidth of 8 MHz provides a total capacity of 4.98 Mb/s, which can be divided between one or more electronic service guides (ESG) and various audio/video subchannels with typically around 200–500 kb/s bit stream dedicated for each. The capacity does not depend on the number of carriers (FFT mode), but the selected FFT affects the Doppler shift tolerance, i.e., the maximum speed of the terminal. As a comparison, the parameter set of 16-QAM, GI 1/32, CR 7/8, and channel bandwidth of 8 MHz provides a total capacity of 21.1 Mb/s. It should be noted, though, that the latter parameter set is not practical due to the clearly increased *C/N* requirement. The relation between the radio parameter values, Doppler shift tolerance, and capacity can be investigated more thoroughly in [Dvb06].

The DVB-H implementation guidelines show the capacity values for the typical parameter settings (Table 10.1).

As an example of Table 10.1, if the QPSK modulation is used with the CR of 1/2 and GI of 1/4, the bandwidth of 6 MHz provides 3.73 Mb/s. Now, if MPE-FEC rate of 3/4 is selected, the usable bit rate is reduced accordingly, resulting $3/4 \times 3.73$ Mb/s, which yields about 2.8 Mb/s. The final capacity values for all the MPE-FEC rate values of 1/2, 2/3, 3/4, and 5/6 can be obtained equally by observing the values of Table 10.1.

The value of, e.g., 2.8 Mb/s can be divided for the time-sliced channels, e.g., by offering 250 kb/s combined A/V stream for each one (note that the value can be selected differently for each channel). The ESG requires its own piece of the capacity, and there can be one or several ESGs defined into a single band. The capacity value depends on the amount of the information ESG delivers. Assuming that the ESG reserves 300 kb/s and there is only one ESG defined into the frequency band, the total amount of A/V channels is (2.8 Mb/s–300 kb/s)/250 kb/s = 10 channels.

As Table 10.1 shows, the 64-QAM mode with 7/8 CR and 1/32 GI provides more than 30 Mb/s for the frequency band. In practice, this parameter setting is very sensible for the errors and requires nearly ideal radio conditions. If SFN in used, this GI value is furthermore very limiting. Moreover, it has been noted that 64-QAM has sensible modulation for the errors and thus limits the usable coverage area. In practice, thus, QPSK and 16-QAM modulations are the recommended ones, whereas 64-QAM could be used, e.g., in limited slow pedestrian environments like shopping centers as it offers superior capacity over the other modulation schemes.

In addition to the example presented above, there are also other items to be considered in the radio capacity calculations. For example, 13 high-quality TV channels can be offered in 6 MHz band when 16-QAM, GI of $^1/_4$, and CR of 1/2 are used offering a total of 7.6 Mb/s. The continuous SI/PSI table delivery also requires one part of the capacity. In this example, 7% should be reserved for this information. Each service also requires a certain proportion of the IP headers, which reduces the usable bit rate accordingly per channel. Figure 10.7 shows the division of different capacity requirements assuming that video stream uses CIF or QCIF class of resolution with H.264 video coding and 15 frames per second, requiring 384 kb/s. The audio stream can be created by using AAC or AAC +, resulting 32 or 64 kb/s streams.

Table 10.1 The summary of DVB-H bit rates in function of the parameter values as presented in [Dvb06]

Mode	Code rate	GI = 1/4			GI = 1/8			GI = 1/16			GI = 1/32		
		6 MHz	7 MHz	8 MHz	6 MHz	7 MHz	8 MHz	6 MHz	7 MHz	8 MHz	6 MHz	7 MHz	8 MHz
QPSK	1/2	3.73	4.35	4.98	4.14	4.83	5.53	4.39	5.12	5.85	4.52	5.27	6.03
	2/3	4.97	5.80	6.64	5.52	6.45	7.37	5.85	6.83	7.81	6.03	7.03	8.04
	3/4	5.59	6.53	7.46	6.22	7.25	8.29	6.58	7.68	8.78	6.78	7.91	9.05
	5/6	6.22	7.25	8.29	6.91	8.06	9.22	7.31	8.53	9.76	7.54	8.79	10.05
	7/8	6.53	7.62	8.71	7.25	8.46	9.68	7.68	8.96	10.25	7.91	9.23	10.56
16-QAM	1/2	7.46	8.70	9.95	8.29	9.67	11.06	8.78	10.24	11.71	9.04	10.55	12.06
	2/3	9.95	11.61	13.27	11.05	12.90	14.75	11.70	13.66	15.61	12.06	14.07	16.09
	3/4	11.19	13.06	14.93	12.44	14.51	16.59	13.17	15.36	17.56	13.57	15.83	18.10
	5/6	12.44	14.51	16.59	13.82	16.12	18.43	14.63	17.07	19.52	15.08	17.59	20.11
	7/8	13.06	15.24	17.42	14.51	16.93	19.35	15.36	17.93	20.49	15.83	18.47	21.11
64-QAM	1/2	11.19	13.06	14.93	12.44	14.51	16.59	13.17	15.36	17.56	13.57	15.83	18.10
	2/3	14.92	17.41	19.91	16.58	19.35	22.12	17.56	20.49	23.42	18.09	21.11	24.13
	3/4	16.79	19.59	22.39	18.66	21.77	24.88	19.76	23.05	26.35	20.35	23.75	27.14
	5/6	18.66	21.77	24.88	20.73	24.19	27.65	21.95	25.61	29.27	22.62	26.39	30.16
	7/8	19.59	22.86	26.13	21.77	25.40	29.03	23.05	26.89	30.74	23.75	27.71	31.67

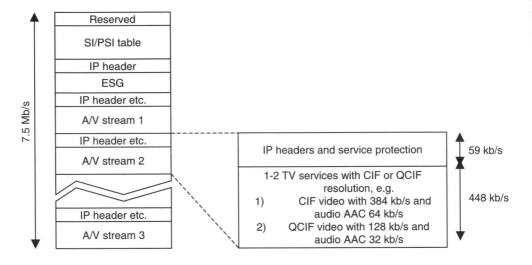

Figure 10.7   An example of the capacity division in the radio interface

10.3   Link Budget

There exist various models for the estimation of the cell radius of mobile communications networks. Some of the possibilities are as follows:

- Okamura–Hata model
- Free space model
- Epstein–Peterson model
- Longley and Rice model
- ITU-R P.370 (CCIR-370) model
- L&S GEG model
- L&S VHF/UHF model
- GE89 model
- ITU-R P.1546 model
- IRT 2D and 3D models
- L&S FCC Point-to-point FM model
- COST207 TY 6tap
- CRC.

As an example of the achieved DVB-H quality, CR of $^{1}/_{2}$, MPE-FEC rate of $^{3}/_{4}$, and 16-QAM modulation would result in 17.5 dB C/N requirement, and a total channel capacity of 6.2 Mb/s. Using the basic SFN, this combination would be possible to use, e.g., for 1 ESG (about 200–300 kb/s) and for 10–12 high-quality A/V channels of about 450 kb/s each, or for about 20 good-quality A/V channels of about 250 kb/s each.

In the initial phase of the planning, e.g., an Okumura–Hata-based rough estimation about the needed amount of sites can be carried out using the respective area correction. For the estimation of the cell size (radius), a link budget is a proper tool. In this specific case, the link budget shown in Table 10.2 can be created.

In the link budget presented in Table 10.2, there are some values that can be assumed as fixed. The noise floor is slightly temperature dependent, but for the link budget purposes a typical temperature value of 25 °C can be used. The radiating power is a result of the transmitter power level, the respective cable, jumper and power splitter losses, and antenna gain. The area location probability level results an additional value that can be reduced from the link budget. In highly available service areas, this loss is logically greater than in low availability network. For the building attenuation value, a typical average value can be used. If the

Table 10.2 Typical DVB-H link budget

General parameters		
Frequency	f	680.0 MHz
Noise floor for 6 MHz bandwidth	P_n	− 106.4 dBm
RX noise figure	F	5.2 dB
TX		
Transmitter output power	P_{TX}	2400.0 W
Transmitter output power	P_{TX}	63.8 dBm
Cable and connector loss	L_{cc}	3.0 dB
Power splitter loss	L_{ps}	3.0 dB
Antenna gain	G_{TX}	13.1 dBi
Antenna gain	G_{TX}	11.0 dBd
Eff. isotropic radiating power	EIRP	70.9 dBm
	EIRP	12308.7 W
Eff. radiating power	ERP	68.8 dBm
	ERP	7502.6 W
RX		
Min C/N for the used mode	$(C/N)_{min}$	17.5 dB
Sensitivity	P_{RXmin}	− 83.7 dBm
Antenna gain, isotropic ref	G_{RX}	− 7.3 dBi
Antenna gain, 1/2 wavelength dipole	G_{RX}	− 5.2 dBd
Isotropic power	P_i	− 76.4 dBm
Location variation for 95% area prob	L_{lv}	5.3 dB
Building loss	L_b	14.0 dB
GSM filter loss	L_{GSM}	0.0 dB
Min required received power outdoors	$P_{min(out)}$	− 71.1 dBm
Min required received power indoors	$P_{min(in)}$	− 57.1 dBm
Min required field strength outdoors	$E_{min(out)}$	62.8 dBuV/m
Min required field strength indoors	$E_{min(in)}$	76.8 dBuV/m
Maximum path loss, outdoors	$L_{pl(out)}$	142.0 dB
Maximum path loss, indoors	$L_{pl(in)}$	128.0 dB

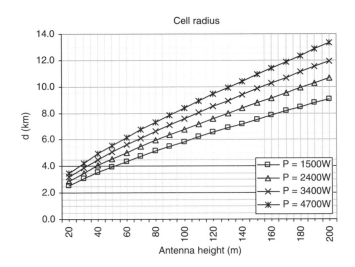

Figure 10.8 The cell range calculated with the Okumura–Hata model for the large city, with varying transmitter power levels

operator wants to make sure the best estimation of this value, comparative field tests can be carried out indoors and outdoors. It should be noted, though, that in order to achieve sufficient statistical accuracy, several tens of measurement cases should be carried out.

Figure 10.8 presents the estimated cell range of the example that is calculated with the large city model and by varying the transmitter antenna height and power level. As can be noted, the antenna height has major impact on the cell radius compared to the transmitter power level.

Figure 10.9 shows the main elements that can be taken into account in the link budget estimations.

10.3.1 Link Budget and Coverage Building Strategies

Link budget in nutshell is the maximum radio signal attenuation between transmitter output and receiver input over which the radio connection works with defined (adequate) quality. Table 10.3 shows a link budget calculation for three transmitter sizes. In reality, there is almost continuous spectrum of transmitter powers available.

The sensitivity figure $-90\,$dBm corresponds to modulation QPSK $^1/_2$ with MPE-FEC, so approximately 4 Mb/s signal. Some operators might want to use higher bit rate with 16-QAM modulation, but then about 6 dB needs to be given up in link budget. Observing the maximum radio path losses from the example of

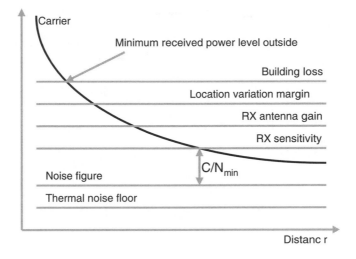

Figure 10.9   The main principle of the DVB-H link budget

configurations in the link budget table, the three classes have values of 158, 145, and 133 dB, respectively, for outdoor coverage. Figure 10.10 displays some useful propagation models for DVB-H purposes, i.e., the Okumura–Hata formula for urban and open terrain environments, and two curves based on [Erc99]. The latter ones are named "EM" in the figure. The upper one of them is for flat terrain that has fairly sparse vegetation, especially trees. The next one is a medium case, when the terrain is either somewhat hilly, or there is substantial amount of trees. Conventionally, it has been assumed that suburban areas have 10–15 dB lower path loss compared to urban Okumura model and it conforms fairly well to the more thorough newer measurements in [Erc99].

The lowest class corresponds to physically a fairly small configuration that could be added to almost any cellular site. The equipment could be a transmitter or a repeater. This option requires that the network operator has fairly good possibilities

Table 10.3   Example of parallel link budgets

Transmitter RF power	W	1000	100	10
	dBW	30	20	10
	dBm	60	50	40
Antenna gain including cable loss	dBi	15	12	10
	dBd	13	10	8
ERP	kW	20	1.0	0.06
Receiver antenna gain	dB	− 7	− 7	− 7
Receiver sensitivity	dBm	− 90	− 90	− 90
Maximum path loss	dB	158	145	133

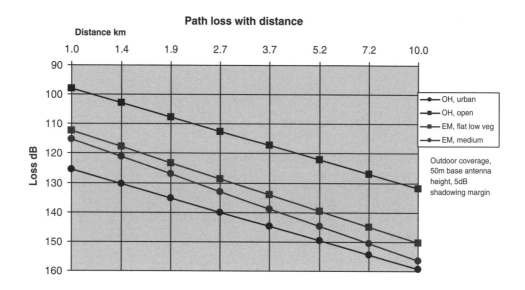

Figure 10.10   Path loss analysis

to arrange cost-efficient transport to the big number of sites needed if they are transmitters. If they are repeaters, then there are other complications with antenna installation. The coverage of one transmitter or repeater is hardly 1 km at above-rooftops antenna installation; in fact in urban centers, almost all base station sites need one. The coverage of one such small equipment is less than a normal cellular BS, but the cellular networks at busy areas typically have quite high overlap coverage. Outside the central areas, where cellular sites get sparser but building height is still several floors, if the sites are more than 1 km apart there may be coverage challenges by using only these small systems.

The second case is a 100 W transmitter, which enables 145 dB path loss outdoors and effective 125–135 dB indoors. This gives 1–2 km indoor range at urban and 3–4 km range at suburban areas. The antenna height at the path loss figure is 50 m and the availability of this kind of sites becomes a question. If the power of the transmitter is some hundreds of Watts, and if the city structure allows finding such a transmitter site every 3–4 km at urban and 5–7 km at suburban areas, this could be a viable option. The transmission arrangements get quite easier compared to case one since many of the sites probably have a fiber optic transmission available.

Going beyond 1 kW transmitter power and 100 m antenna heights is traditional high power TV transmitter technology. This kind of site might give indoor coverage of several kilometers, but still hardly beyond 10–15 km at urban areas. Anyway the viability of these sites needs to be studied and measured case by case. Quite often, powerful TV sites are farther than this from the city centers, so it is not possible to rely only on this technology even if it somewhere would be a good choice.

It seems that no option alone for coverage building is the best strategy. Normally with radio coverage building, the rule is that the bigger the cheaper coverage, but due to constraints of using high sites, "one size fits all" strategy may not be the best one.

10.4 Coverage Area Calculations

10.4.1 Okumura–Hata

According to the link budget presented in Section 10.3, the outdoor reception of this specific case with 2400 W transmitter yields a successful reception when the radio path loss is equal to or less than 142.0 dB.

The expected radius of the cell can be estimated with Okumura–Hata model. It is especially suitable for the relatively reduced broadcast environment as the model is most accurate in the transmitter antenna height of 30–200 m, and the radius does not exceed 20 km.

The Okumura–Hata model can be applied in order to obtain the cell radius (unit in kilometers), for example, in large city type in the following way:

$$L(\text{dB}) = 69.55 + 26.16\lg(f) - 13.82\lg(h_{\text{BS}}) - a(h_{\text{MS}})$$
$$+ [44.9 - 6.55\lg(h_{\text{BS}})]\lg(d)$$

$$a(h_{\text{MS}})_{\text{LC1}} = 8.29[\lg(1.54h_{\text{MS}})]^2 - 1.10, \quad f \leq 200\,\text{MHz}$$
$$a(h_{\text{MS}})_{\text{LC2}} = 3.2[\lg(11.75h_{\text{MS}})]^2 - 4.97, \quad f \geq 400\,\text{MHz}$$

$$d = 10^{\left(\frac{L(\text{dB}) - [69.55 + 26.16\lg(f) - 13.82\lg(h_{\text{BS}}) - a(h_{\text{MS}})]}{44.9 - 6.55\lg(h_{\text{BS}})}\right)}$$

For the rest of the environments, please refer to, e.g., original report presented in [Hat80].

10.4.2 ITU-R P.1546 Model

Even the Okumura–Hata model has been extended for the typical DVB-H as for the larger cell sizes, International Telecommunications Union (ITU) has identified a need for the propagation models that suit in basically all the broadcast cases. The latest ITU-R P.1546 model fits basically all the environments where DVB-H is used [ITU07].

The model is based on the predefined curves for the frequency range of 30–3000 MHz and for maximum antenna heights of 3000 m from the surrounding ground level. The model is valid for the terminal distances of 1–1000 km from the base station over terrestrial and sea levels, or in the combination of these.

If the investigated frequency and antenna height do not coincide with the predefined curves, the correct values can be obtained by interpolating or extrapolating the predefined values. The presented curves represent field strength values for 1 kW effective radiated power level (ERP), and the curves have been produced for 100 MHz, 600 MHz, and 2 GHz. The curves are based on the empirical studies about the propagation in certain conditions. In addition to the graphical curve format, the values can also be obtained in tabulated numerical format.

It can be assumed that the basic and extended version of Okumura–Hata as well as ITU-R P.1546 models provide a good first-hand estimate for the DVB-H coverage areas and respective capacity and quality levels in the initial network planning phase. These models have been designed for environments with antenna heights and cell distances that fall into the typical assumptions of DVB-H networks. The [Mil06] identifies several other models, including ray-tracing type of estimates for the dense city centers. These models require more detailed map data with respective height and cluster attenuations. In the most advanced versions, a vector-based 3D map is needed. It logically has a cost effect on the planning but increases considerably the accuracy of the coverage estimate. It can further be enhanced via the reference measurements by tuning the estimate respectively. Even with the cost effect, it would be thus justified to use 3D models in the advanced phase of the radio network planning.

10.4.3   Building Penetration Loss

When designing the indoor coverage, the respective building loss should be taken into account. The loss depends on the building type and material. As an example, Figure 10.11 shows a short measurement carried out in outdoor and indoor of a 10-floor hotel building within the coverage area of the investigated DVB-H network.

The building loss of Figure 10.11 can be obtained by calculating the difference of the received power values. The indoor average received power level is − 73.4 dBm with a standard deviation of 5.7 dB. The outdoor values are − 57.3 dBm and 3.9 dB, respectively. The building loss in this specific case is thus 16.1 dB. The value is logical as the building represents relatively heavy construction type, although the rooftop was partially covered by large areas of class resulting in relatively good signal propagation to interior of the building via the diffraction.

In the general case, the building loss should be estimated depending on the overall building type, height, etc. in each environment type. In case of new areas, the best way for the estimation is to carry out sufficient amount of sample measurements for the most typical building types, although for the initial link budget estimations, the average of 12–16 dB could be a good starting point according to these tests.

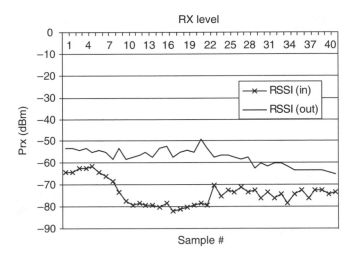

Figure 10.11   An example of short snap-shot type indoor and outdoor measurements carried out in the DVB-H trial network coverage area

10.4.4   Case Example of the Dense Urban/Urban Environment

The coverage area of a single DVB-H transmitter site depends on the provided capacity and thus on the radio parameters that provide the required bit pipe. Logically, higher total capacity demand in the given bandwidth results in smaller cell sizes.

The environment has a big impact on the radio wave propagation. According to the Okumura–Hata prediction model, the dense urban area attenuates the signal considerably compared to the other environment types. This section shows an example of the coverage planning in Mexico City, which is considered as one of the largest urban and dense urban environments. Despite the challenges of the presented environment type, the definite advantage of DVB-H can be seen most clearly in this case as a big amount of potential customers can be served in reduced area. In Mexico City alone, the estimated total population is about 20–25 million from which the number of the potential customers can be estimated to be considerable.

10.4.4.1   The Planned Environment

Mexico City is situated 2.2 km of height from the sea level, and it is surrounded by the mountains with about 3 km of height compared to the sea level. Figure 10.12 shows an overview of the city. As can be observed, the area consists of urban buildings in large area.

Figure 10.13 shows the area cluster type of Mexico City. As can be seen, the dense urban and urban type is very large with the respective cluster type proportion of roughly 1000 km^2.

Figure 10.12   An overview of Mexico City. The photo shows the densest area. The city is in general tightly built and large in size

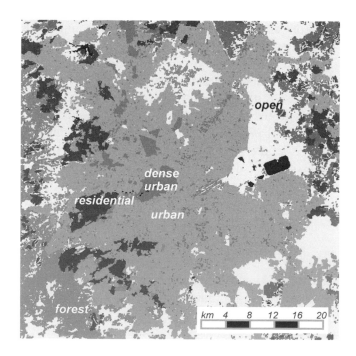

Figure 10.13   The area type, i.e., cluster map of Mexico City. This map shows the area of about 50×50 km in which half is urban and dense urban area

10.4.4.2   Okumura–Hata Analysis

Sections 10.3 and 10.4.1 can be used as a basis for the following analysis.

It is clear that higher the antenna is located and higher the transmitter power is, lower is the transmitter site number. In practice, though, it is not always possible to obtain the site locations and antenna heights in the technically best locations. The access for already existing towers might be limited as well as the available heights in the towers, and rooftops might be challenging to obtain. In many cases, the antenna height is limited to 20–30 m. In some cases, it could be possible to obtain a higher antenna location in broadcast tower near the city center, or even better, in sky-scraper's rooftop in the downtown area.

On the other hand, the transmitter type is important to select correctly. In case of high-power transmitter type, the respective power consumption increases. It can be estimated that the power consumption might be about six times the produced power level fed to the antenna cable. Also the complexity rises among the higher power levels and, e.g., liquid cooling is needed instead of air cooling, affecting the main-tenance. There might be limitations for the highest power classes when selecting the optimal power levels.

In this analysis, a 2400 W transmitter type was selected due to the above-mentioned reasons. According to the initial link budget calculation presented in Table 10.2, it provides a cell radius of 3–4 km with the transmitter antenna installed to 20–40 m of height. As a comparison, the antenna height of 100 m provides about 7 km cell radius, and 200 m antenna yields about 10–11 km radius. If possible to install, one or two high antenna locations would provide a good basic coverage in the city area while the sites with lower antenna heights fills the rest of the area.

If only the urban and dense urban areas are to be covered, the following theoretical coverage map can be created by selecting two sites with antenna height of 200 m and cell radius of 11 km, and the rest of the sites could use antenna heights of 30 m, which provides a cell radius of about 3.5 km (Figure 10.14). The map represents the Okumura–Hata estimated coverage for the outdoor assuming that the sites can be selected without restrictions.

The drawback of the earlier presented coverage plan is that the prediction is inaccurate depending on the actual terrain type. Nevertheless, it gives an idea about the rough cell number. In any case, this analysis shows the importance of the antenna height as with only two high antenna locations; it looks possible to cover about half of the given area while the low antenna locations produce a need for about 15 sites.

10.4.4.3   ITU-R 1546 Analysis

The ITU-R 1546 model, using the version 3, can be used assuming that the frequency is 680 MHz as in the previous Okumura–Hata analysis. The ITU-R

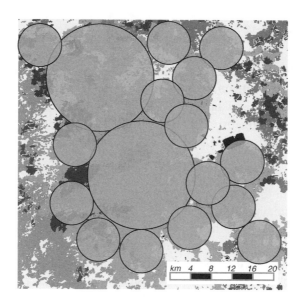

Figure 10.14 Theoretical coverage plan when using Okumura–Hata model for the cell radius estimation

model's curves do not coincide with this frequency, so we need to interpolate the investigated frequency by using the 600 MHz and 2 GHz curves.

The step-by-step explanation of the model can be found in [ITU07]. Let us take an example of the mountain site as the ITU-R model is especially useful for this type of analysis. Figure 10.15 presents the profile of the site.

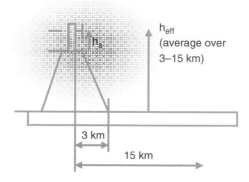

Figure 10.15 The terrain profile of the mountain site presented in the example. The mountain peak is 800 m above the surrounding terrain level

In this specific example, the mountain peak is in 800 m from the surrounding terrain level. In addition, the antenna is located at 60 m height in the tower, resulting a final antenna height $h_a = 860$ m.

Now, let us observe the land path values. For the paths shorter than 15 km, and where no terrain information is available, the effective antenna height is $h_1 = h_a$ for the distances of less than 3 km, and for the distances between 3 and 15 km, the effective antenna height is

$$h_1 = h_a + (h_{\text{eff}} - h_a)\frac{(d-3)}{12}$$

As an example, for the distance of 1 km, the effective antenna height is directly the antenna height in the tower, i.e., 60 m, whereas at 10 km, the antenna height is 526.7 m according to the above-mentioned formula. In the greater distance of 15 km, the effective antenna height is the maximum, i.e., the mountain and tower installation height (860 m).

The curves present the situation for the "virtual" mobile antenna height of R (in meters). This R thus presents an area cluster weighted antenna height. The values are assumed as 10 m for the suburban area, 20 m for the urban area, and 30 m for the dense urban area in land. For the water, 10 m is assumed for all the cases. In order to take the arriving angle into account, a modified clutter height R' should be calculated, d being the distance in kilometers:

$$R' = \frac{1000dR - 15h_1}{1000d - 15}$$

As an example, for the dense urban cluster with R of 30 m, transmitter antenna being at 860 m of height, the modified clutter height is 28.8 m.

If the mobile is within urban area type and its antenna height is 1.5 m, which is less than the R', the correction is $6.03 - J(v)$, where $J(v)$ is presented as (frequency f being in MHz):

$$J(v) = \left[6.9 + 20\log\left(\sqrt{(v-0.1)^2 + 1} + v - 0.1\right)\right]$$
$$v = K_{\text{nu}}\sqrt{h_{\text{dif}}\theta_{\text{clut}}}$$
$$h_{\text{dif}}(\text{m}) = R' - h_2$$
$$\theta_{\text{clut}}(^\circ) = \arctan(h_{\text{dif}}/27)$$
$$K_{h2} = 3.2 + 6.2\log(f)$$
$$K_{\text{nu}} = 0.0108\sqrt{f}$$

The antenna height is another parameter that has to be interpolated as the closest height that the ITU-R model defines occurs in 600 and 1200 m.

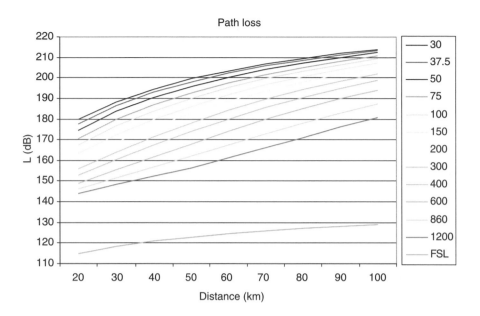

Figure 10.16   The interpolated path loss (L) curve for the 860 m antenna height when the frequency is 680 MHz. The graph also shows the values for the predefined antenna heights. FSL represents the free space loss curve

The interpolation of the frequency and antenna height results in Figure 10.16, with the predefined antenna height curves calculated and presented in the same graph as the mountain site case with 860 m of the antenna height.

It is worth noting, though, that the coverage area of the mountain site towards the center area depends on the antenna direction as there are various smaller terrain peaks between the site and center as can be seen in Figure 10.17.

10.4.4.4   Planning Tool Analysis

In order to compare the theoretical Okumura–Hata approach with the more realistic methods, Nokia NetAct Planner was used as a basis for more in-depth analysis of the coverage planning. The tool consists of the digital maps of Mexico City, with respective clutter data. A total of seven sites were selected for the analysis based on the practical site considerations, i.e., the selected sites could possibly be real candidates with realistic antenna heights.

The NetAct consists of several propagation models. Extended Okumura–Hata prediction model with respective digital cluster maps was used in the analysis. The initial parameter tuning for the DVB-H plan was made based on the estimated local clutter attenuation factors. The final clutter values and other propagation model parameterization for the estimated coverage area should be adjusted by carrying out field tests as each area type differs from the others.

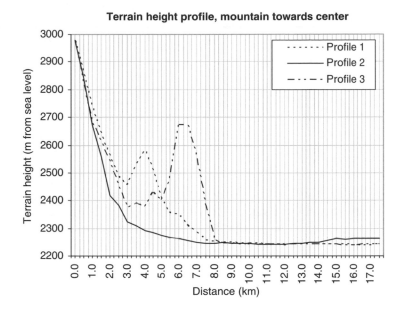

Figure 10.17   The terrain height profile of the investigated mountain site

The coverage map was plotted for outdoor and indoor environments based on the previously used DVB-H link budget, taking into account the relevant parameters (bandwidth, modulation scheme, CR, MPE-FEC rate and receiver antenna gain). The same 2400 W transmitter type was used in all the sites like in previous analysis, with EIRP of about 69–71 dBm, depending on the site configuration, i.e., cable lengths and losses. In this analysis, two- or three-directional antennas with the horizontal beam width of 65° and vertical beam of 27° were used, with respective antenna gain of 13.1 dBi. In some cases of high antenna installations, a slight antenna element down-tilting was used in order to optimize the coverage.

Other essential global parameters for the link budget were CR $^1/_2$, MPE-FEC rate $^3/_4$, radio channel TU6 (with 30 Hz Doppler), channel bandwidth 6 MHz, and 680 MHz operating frequency. The average building penetration loss was estimated to be 14 dB. With the area location probability of 95%, the link budget yields a minimum requirement of 71.1 dBm for the received power level in this specific case.

Table 10.4 presents the selected sites with the antenna height h_{ant}, direction (degrees), down-tilt (DT), and the final EIRP (dBm) values. The EIRP shown in the table includes the transmitter filter, cable, connector, and power splitter loss.

The coverage plots indicate the functional areas for 16-QAM in outdoor and indoor environments (Figure 10.18). As a reference, also QPSK outdoor coverage is presented. In the following coverage maps, the raster size is 5 × 5 km (Figures 10.19 and 10.20).

Table 10.4 The DVB-H site configuration

Site	TX P (W)	h_{ant} (m)	Degree	DT	EIRP
Tres Padres	2400	60	140/220	2/2	70.5
WTC	2400	190	0/150/240	2/2/2	69.3
Iztapalapa	2400	30	0/120/240	0/0/0	69.3
Santa Fe	2400	20	0/120/240	0/0/0	69.5
Tlalpan	2400	20	330/90	0/0	71.3
Vallejo	2400	30	0/120/240	0/0/0	69.3
Azteca	2400	30	120/240	0/0	71.1

The cell size estimation obtained by calculating the pure Okumura–Hata prediction model has relatively good average correlation with the results that can be obtained with NetAct Planner when the general limits of the models are taken into account.

The results of the NetAct Planner show that the used prediction takes well into account the terrain heights and cluster types, which would be challenging to do with using only theoretical Okumura–Hata approach. When sufficiently good line of sight is found in the planned sector, the useful cell size may be considerably better that obtained with the use of Okumura–Hata.

More sites are obviously needed if the same area has to be covered as shown in Okumura–Hata analysis in Figure 10.14. By observing Figure 10.20, about four to six additional sites might be necessary for the full coverage. Okumura–Hata

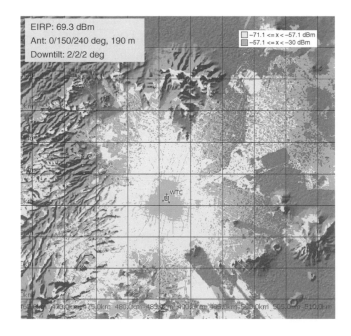

Figure 10.18 Example of the great antenna height

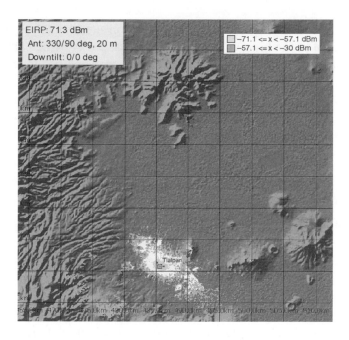

Figure 10.19 Example of the coverage when the antenna is located low

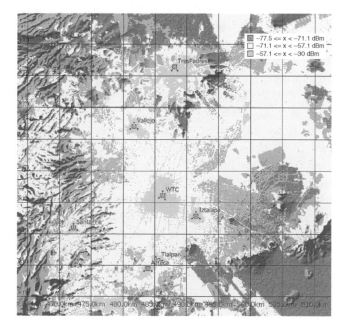

Figure 10.20 The complete DVB-H network coverage when all the sites are activated

estimated well the relatively low antenna installation sites, but the model estimated the coverage area of the high WTC site in pessimistic way, which affects the final estimation of the sites. The local adjustment of the propagation model for different environments is thus important in order to estimate the cell count close to the reality as much as possible.

It is worth noting that especially the indoor coverage in selected areas requires the use of repeater type of solution, e.g., in shopping centers and other centralized locations, where the potential customers are typically using the service.

10.4.4.5 Conclusions

The results of this case analysis show that the theoretical Okumura–Hata prediction model with DVB-H link budget gives a good first-hand estimate about the cell sizes and thus about the needed amount of sites in the planned area. Taking into account the characteristics of the model, this method can be applied especially in the initial phase of the network planning.

Due to the restrictions of the Okumura–Hata ranges as the antenna height and maximum estimated cell radius are considered, the methodology applies for the relatively low radiating power levels. When the cell radius exceeds the maximum predictable value of 20 km, the model becomes infeasible and adjusted models should be used. Especially for the high antenna locations, one of the most logical models at the moment is the ITU-R P.1546. On the other side, the practical power levels are limited due to the EMC and human exposure regulation resulting in sufficiently small cell ranges in order to be estimated with Okumura–Hata in major part of the cases in urban areas.

The advanced planning tool with respective digital maps including the terrain height and correct cluster attenuation information is essential in the detailed network planning. It is also worth noting that the predictions presented in this analysis gives indication only about the coverage areas. Especially in the case of over-dimensioned SFN, the correct balancing of the FFT size and GI values is important in order to avoid too high level of the possible inter-symbol interferences in large SFN areas.

The coverage estimation presented in this section can be used in the first phase of the DVB-H radio network planning for the rough estimation of the transmitter sites. As the clutter types vary in practice, more detailed prediction estimations with respective model tuning via the field tests are thus needed in the following phases.

10.4.5 Case Example of the Suburban Environment

The investigation of the DVB-H radio propagation was carried out in low-power suburban area type as described in [Pen08]. The area consisted of light vegetation, one- or two-floor residential single-family houses and relatively open areas between the houses and motorways. The center of the radio network was a 200-W transmitter with a DVB-T/H modulator. The power level that was delivered to the antenna cable after the transmitter's filter was about 175 W that results in radiating power (EIRP) of 62 dBm.

Two directional antennas (13.1 dBi, 65/27° beam width in horizontal/vertical plane) were installed on top of each others to create narrower vertical beams. The antennas were connected to the transmitter with a power splitter. The antenna height from the surrounding ground level was about 60 m.

A plot of the estimated coverage area was produced by using the Nokia NetAct Planner. The aim was to concentrate only on the coverage area. Figure 10.21 represents the estimated coverage area for QPSK and 16-QAM cases with the CR of $^1/_2$ and MPE-FEC rate of $^3/_4$. The Okumura–Hata-based propagation model was used with a digital map that contains the elevation data and cluster-type information.

Based on the coverage plots, the test route was selected accordingly. The coverage plots correlated with the results obtained via the drive tests. The grid size is 1.6 km.

An area location probability of 95% was applied in this analysis, resulting in the received power level references of -77.9 dBm for QPSK and -71.1 dBm for 16-QAM. The dark grey color represents the outdoor coverage estimation of DVB-H, and the other colors represent the incremental of the performance by 1 dB per each color. A grid size of 1 mile or 1.6 km is used for all the maps.

The results show that in the relatively small scale, the cell radius of DVB-H can be estimated sufficiently accurately by using Okumura–Hata model. It is worth noting, though, that the model's parameter tuning should be adjusted accordingly before it can be used in the most accurate way. Nevertheless, the model gives a sufficiently

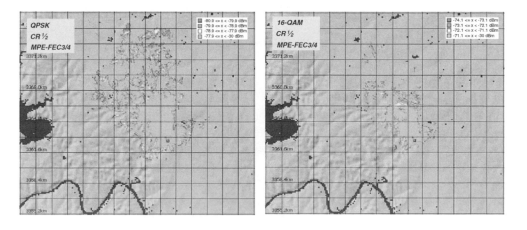

Figure 10.21   Estimated coverage area of suburban case study

good understanding about the expected cell sizes as such, as well as when the model is used as a base in the network planning programs with digital maps.

10.5 Trade-offs Between the Parameters

10.5.1 Coverage and Capacity

The coverage and capacity have a clear trade-off on DVB-H radio network. More capacity can be obtained by using more efficient modulation schemes, i.e., 16-QAM provides about double capacity over QPSK, and 64-QAM provides respectively more, but each time the coverage reduces as the more capacity and efficient modulation schemes require higher C/N. Also the CR affects the capacity. Nevertheless, more capacity can be squeezed out form the radio interface via lighter CR modes; they require more C/N reducing the useful coverage area. MPE-FEC reduces the capacity accordingly, but enhances the C/N performance.

Figure 10.22 shows an example about the effect of the balancing of capacity and coverage area when the antenna height is 100 m, 1 kW EIRP is used and antenna gain of 10 dBi is applied. The channel bandwidth is 8 MHz in this specific case.

10.5.2 Trade-off Between FFT and SFN Size

Reception of the signal of SFN requires that the signals from different transmitters arrive at the receiver in a certain time window. All the transmitters of the SFN

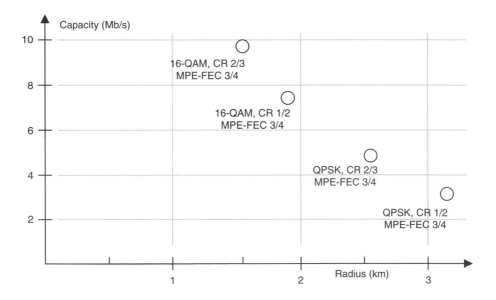

Figure 10.22 An example of the DVB-H trade-off of capacity and coverage area

transmit exactly the same signal, and they are synchronized. The OFDM symbols are so long that a GI has been inserted in between the OFDM symbols in order to allow different propagation times from the transmitters to the receiver. The longest GIs are $^1/_4$ of the data symbol duration, so they eat maximum of 20% of the time. In fact, the guard period (GP) is not "empty" due to transmitter implementation considerations, but it is filled by extending the data symbol by copying its content over the GP. Seen from the receiver, two transmitters in the SFN must never be geographically farther away from each other than the distance determined by velocity of radio signal, i.e., speed of light, and GI. Geometrically, this implies that the transmitters must reside within a circle; diameter of it being speed of light multiplied by the GI. GP values can be chosen at the range 7–256 µs. Table 10.5 gives the diameter of the SFN with different FFT and GP values.

The lowest FFT values, 2K and 4K, give rather small values for the SFN size, regarding the allotment frequency planning of the Geneva 06 plan for the digital television networks. Allotment is frequency assignment for an area, measuring typically 100 km or less. The area can be served by one high-power transmitter or by a number of smaller operated in SFN mode. In realistic SFN networks, the 8K mode is preferable, and 4K mode only if very fast-moving terminals are served. This is another trade-off, which is described later.

The receiver may be located outside or inside the circle of the set of transmitters, and the location of the terminal is actually irrelevant (Figure 10.23). Thus the actual service area of the SFN may be somewhat larger, depending on the service range of the transmitters. If the SFN is implemented with great number of small transmitters, then the service area extends very little outside the circle, but if the transmitters are larger, then the service area may extend clearly out of the circle, at countryside, for

Table 10.5 The relationship between GI and SFN transmitter area

FFT mode	GI		SFN diameter (km)
	Proportion	Microseconds	
8K	1/4	224	67.2
	1/8	112	33.6
	1/16	56	16.8
	1/32	28	8.4
4K	1/4	112	33.6
	1/8	56	16.8
	1/16	28	8.4
	1/32	14	4.2
2K	1/4	56	16.8
	1/8	28	8.4
	1/16	14	4.2
	1/32	7	2.1

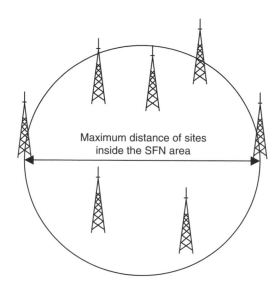

Figure 10.23 The transmitters of the SFN must reside within a circle with the diameter of maximum SFN GI size

example, 10–20 km if we have mid-size transmitters with RF power of several 100 W and antenna heights of $50 - 150$ m.

In networks that have OFDM, e.g., WLAN, WiMAX, and LTE, the OFDM is utilized in order to reach economically viable receiver design with high bit rate signal, but in DVB the OFDM characteristics are also utilized to make SFN transmission possible. Therefore, much longer GPs are applied than the other radio technologies mentioned here, where GPs range from 10 to 20 µs.

10.5.3   Large SFNs and Outlying Transmitters

There is a theoretical possibility to build an infinitely large SFN. If the transmitters of the SFN are small enough, a transmitter that is at the distance of the SFN transmitter area diameter is weak enough not to cause interference to the signal from the nearest serving transmitters. If a uniform network is considered with this technique, it might lead to a requirement that the transmitters are of fairly small size, RF power tens or at most a few 100 W and antenna heights roughly below 50 m. The practical size of SFN is anyway limited to the allotment in question. If there is a need to apply transmitters outside the SFN circle, reducing the power may be a viable option to reach adequate high signal quality. Another method is to adjust the timing of the outlying transmitter to the rest of the transmitters.

10.5.4   Trade-off Between FFT, Modulation, and Maximum Terminal Speed

The FFT size determines the subcarrier spacing in the OFDM signal. The Doppler phenomenon distorts the signal in between the transmitter and receiver. In typical mobile receive conditions, the Doppler effect does not produce a nice clean shift of the signal up or down at the frequency axis, but since the different reflections of the signal arrive to the receiver from various directions, it means that the reflected signal components are shifted up or down in quite a random way, so the subcarriers tend to smear and start overlapping each other when the shift is too large. The frequency-shifted reflected signal components begin to fill the zeros of the OFDM signal, thus reducing the signal-to-interference ratio.

The subcarrier spacing is directly proportional to the FFT value, so the 8K FFT has exactly four times higher tolerance to terminal speed than 2K FFT. The higher modulations require higher signal-to-interference ratio, so the low modulations tolerate higher terminal speeds than high modulations. In Figure 10.24, we see how the performance of reception stays fairly constant with increasing terminal speed, but when approaching the maximum speed limits it degrades suddenly. Figure 10.24 shows the principle of the Doppler frequency behavior for different modes 2K, 4K, and 8K, without MPE-FEC. The terminal speed is presented for the 700 MHz carrier frequency.

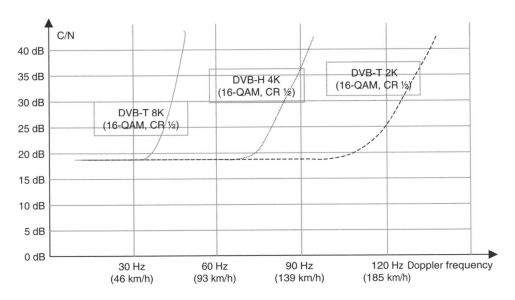

Figure 10.24   The Doppler frequency behavior. Please note that the figure is showing only the principle without exact values, and the more specific values can be found in the ETSI TR 102 377 DVB-H Implementation Guidelines [Dvb06]

For the calculation of the terminal speed limit via the Doppler shift, an example of the CR 2/3, MPE-FEC 5/6, and a frequency of 610 MHz is presented. The frequency 610 MHz corresponds to about 50 cm of wavelength according to the formula $f = c/\lambda$. The Doppler frequency $= 33.3$ (120 km/h)/$0.50 = 67$ Hz. In this case, for the urban link budget, choose 60 Hz Doppler.

10.5.5 Trade-off Between Maximum Terminal Speed and Carrier Frequency

The Doppler shift is directly proportional to the carrier frequency, whereas the signal bandwidth of course stays constant regardless the RF; so the higher the frequency, the higher the Doppler shift, and therefore higher frequencies suffer more from the terminal speed. The effect is linear with RF, 25% higher frequency causes 25% lower maximum speed.

10.5.6 Trade-off Between Bit Rate and Coverage

As with many other radio systems that have adaptive modulation and coding, there is a trade-off between service bit rate and coverage. The trade-off is rooted from the Shannon formulas, and it is a law of the nature. We can only adapt to it. Figure 10.25 displays the relative coverage of different bit rates. The sensitivity is found, e.g., in ETSI EN 300 744 Annex A, simulated system performance table. The sensitivity

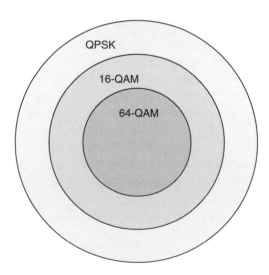

Figure 10.25 The principle of the relationship of coverage areas of different modulations

decreases to roughly 6 dB for modulation increase of QPSK to 16-QAM or 16-QAM to 64-QAM. An increase of 6 dB in link budget corresponds to a coverage increase of approximately twofold. The coverage depends on signal level decrease with distance from transmitter. The 64-QAM has four times the bit rate of QPSK, and it is somewhat a coincidence that the propagation formulas determine the coverage area of the QPSK quite close to four times larger. But this relationship of coverage and bit rate leads to a simple rule, doubling the capacity also doubles the number of transmitters needed if similar transmitters are applied.

10.5.7 Trade-offs and Network Characteristics

As described in the preceding paragraphs, the parameters that can be set by the network designer, or the network operator, have an affect on the network characteristics. The main service characteristics of the network can be seen as coverage, bit rate, and maximum terminal speed (Figure 10.26). The parameters that can be modified are, for example, transmitter size, modulation and coding scheme (MCS), FFT value, and GI. These parameters tend to move the service characteristics within a triangle; when one characteristic is emphasized, or maximized, there is a tendency that the other two need to give up.

Low MCS gives good coverage and good speed performance, but low bit rate. The higher the transmitter power, the better the coverage and bit rate, but then the SFN size needs to be fairly large, which requires high FFT. In practice, the SFN size is tied to the existing frequency plan – in Europe, Africa, and Middle-East the Geneva 06 plan. In case when the frequency planning would be at the hands of the operator, the determination of the SFN areas would be possible for the network operator. For the large SFNs, long GI is needed, which in turn reduces the bit rate.

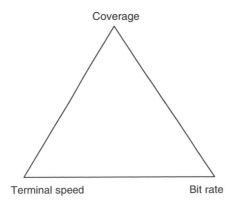

Figure 10.26 The relationship between the coverage area, useful bit rate and terminal speed

10.6   Radio Network Measurements and Analysis

As the core network is IP based, the related, general network analyzers, testers, trouble shooting equipment, etc. In this sense, the standard IP-based core network measurement equipment can be used in normal cases in DVB-H.

There is a variety of DVB-H field measurement equipment available for the radio interface performance revisions. Common DVB-T measurement equipment can also be used for the DVB-H revisions as for the basic functionality. As there are DVB-H-specific functionalities, specialized equipment is also needed in order to investigate, e.g., the MPE-FEC functionality and time slicing.

10.6.1   Terminal as a Measurement Equipment

The field measurement equipment that provides reliable results is essential in the quality verification of DVB-H networks. In addition, sufficiently in-depth analysis of the postprocessed data is important. This section presents a method to collect and analyze the key performance indicators of the DVB-H radio interface, using a mobile device as a measurement and data collection unit.

10.6.1.1   Introduction

The verification of the DVB-H QoS level can be done by carrying out field measurements within the coverage area. The correct ways to obtain the most relevant measurement data, as well as the right interpretation of it, are fundamental for the detailed network planning and optimization.

During the normal operation of the DVB-H network, there are only few possibilities to carry out long-lasting and in-depth measurements. A simple and fast field measurement method based on mobile DVB-H receiver thus provides added value for the operator. The mobile equipment is easy to carry both in outdoor and indoor environment, and it stores sufficiently detailed performance data for the postprocessing.

The measurements are required for the network performance revisions and for the indication of potential problems. As an example, the transmitter site antenna element might move due to the loose mounting, which results in outages in the designed coverage area. The antenna feeder might still remain connected correctly, keeping the reflected power in acceptable level. As there are no alarms triggered in this type of instances, and as the basic DVB-H is a broadcast system without uplink and its related monitoring/alarming system, the most efficient way to verify this kind of fails is to carry out field tests.

The guideline defines the MPE frame error ratio (MFER) as a ratio of the number of residual erroneous frames that cannot be recovered to the total number of the received frames:

$$\text{MFER}(\%) = 100\frac{\text{residual erroneous frames}}{\text{received frames}}$$

There are possibilities to obtain the MFER value by storing the frames during certain time (e.g., 20 s), or as proposed by DVB, at least 100 frames should be collected in order to calculate the MFER with sufficient statistical reliability.

10.6.1.2 Case Study

A test setup with a functional DVB-H transmitter site was utilized in order to investigate the presented field test methodology with respective postprocessing and analysis [Pen08]. The investigated area represents relatively open suburban environment as shown in Figure 10.27.

The methodology was verified by carrying out various field tests mostly in vehicles. There were also static and dynamic pedestrian type of measurements included in the test cases in order to verify the usability of the equipment and methodology for the analysis.

The DVB-H test network consisted of a single 200 W DVB-H transmitter and a basic DVB-H core network. The source data was delivered to the radio interface by capturing real-time television program. The program was converted to DVB-H IP data stream with standard DVB-H encoders. There was a set of three DVB-H channels defined in the same RF, with audio/video bit rates of 128, 256, and 384 kb/s. The number of FEC rows was selected as 256 for MPE-FEC rate of $^1\!/_2$, and 512 for MPE-FEC rate of 2/3. The audio part of the channel was coded with AAC using a total of 64 kb/s for stereophonic sound. The bandwidth was 6 MHz in 701 MHz frequency.

The antenna system consisted of directional antenna panel array with two elements. Each element produces 65° of horizontal beam width and provides a gain of +13.1 dBi.

Figure 10.27 The environment in main lobe of the transmitter antenna

The vertical beam width of the single antenna element was 27°, which was narrowed by locating two antennas on top of each others via a power splitter. Taking into account the loss of cabling, jumpers, connectors, power splitter, and transmitter filter, the radiating power was estimated to be + 62.0 dBm (EIRP) in the main lobe.

The transmitter antenna system was installed on a rooftop with 30 m of height from the ground. The environment consisted of suburban and residential types with LOS (line-of-sight) or nearly LOS in major part of the test route, excluding the back lobe direction of the site, which was non-LOS due to the shadowing of the site building. Each test route consisted of two rounds in the main lobe of the antenna with a minimum received power level of about − 90 dBm. The maximum distance between the antenna system and terminals was about 6.4 km during the drive tests.

If the relevant data can be measured from the radio interface and stored in text format, the method presented in this paper is independent of the terminal type. It is important to notice, though, that the characteristics of the terminal affect the analysis, i.e., the terminal noise factor and the antenna gain (which is normally negative in case of small DVB-H terminals) should be taken into account accordingly.

The terminals were kept in the same position inside the vehicle without external antenna, and the results of each test case were saved in separate text files. The terminal setup is shown in Figure 10.28.

Figure 10.28   The terminal measurement setup

10.6.1.3 Terminal Measurement Principles

The DVB-H parameter set was adjusted according to each test case. The cases included the variation of the CR with the values of $^1/_2$ and 2/3, MPE-FEC rate with the values of $^1/_2$ and 2/3, and interleaving size FFT with the values of 2K, 4K, 8K, in accordance with the Wing TV principles described in [Apa06a], [Apa06b], [Bou06], and [Mil06]. The GI was fixed to $^1/_4$ in each case. The parameter set was tuned for each case, and the audio/video stream was received with all the terminals by driving the test route two consecutive times per parameter setting.

According to the DVB-H implementation guidelines, the target QoS is as follows:

* For the bit error rate after Viterbi (BA), the reception level should comply with at least DVB-H-specific QEF (quasi error free) point 2E10 − 4.
* The frame error rate (FER) should be less than 5%. In case of FER, i.e., DVB-T, this criterion is called FER5, and for the DVB-H-specific MPE-FEC, its name is MFER5.

The field test software of N-92 is capable of collecting the RSSI (received power levels in dBm), FER, and MFER (i.e., FER after MPE-FEC correction) values. In addition, there is possibility to collect information about the packet errors (PE).

The measurement point for the received power level is found after the antenna element and the optional GSM interference filter. In addition, there might be optional external antenna connector implemented in the terminal before the RF reference point. The presence of the filter and antenna connector has thus frequency-dependent loss effect on the measured received power level in the RF point.

Figure 10.29 shows an example of the measurement data file. The example shows four consecutive test results. Each measurement field contains information about

```
 FER:  1 MFER:   0
BB:2.5e-02 BA:1.2e-03
PE:   75
RSSI: -89
 FER:  0 MFER:   0
BB:4.5e-02 BA:1.7e-03
PE:   25
RSSI: -89
 FER:  1 MFER:   0
BB:3.6e-02 BA:8.2e-04
PE:    7
RSSI: -88
 FER:  1 MFER:   1
BB:0.0e+00 BA:0.0e+00
PE:    0
RSSI: -87
```

Figure 10.29 Example of the measured objects with four consecutive results for FER, MFER, BB, BA, PE, and RSSI

the occurred frame error (FER), frame error after MPE-FEC correction (MFER), bit error rate before Viterbi (BB) and after Viterbi (BA), PE, and received power level (RSSI) in dBm. In this case, there was a frame error in the reception of the first measurement sample because the value of FER was "1." The FER value is either "0" for nonerroneous or "1" for erroneous frame. The MPE-FEC procedure could still recover the error in this case, because the MFER parameter is showing a value of "0." The second sample shows that there were no frame errors before or after the MPE-FEC. The third sample again shows frame error that could be corrected with MPE-FEC. The fourth sample shows an error that could not be corrected any more with MPE-FEC. In the latter case, the bit error information could not be calculated either. It seems that in this specific case, the RSSI value of about -87 to $-89\,$dBm has been the limit for the correct reception of the frames with MPE-FEC.

The plain measurement data has to be postprocessed in order to analyze the breaking points for the edge of the performance. According to the first sample of Figure 10.29, the bit error level before Viterbi (BB) was close to the QEF point, i.e., 2.5×10^{-2}. The bit error level after the Viterbi (BA) was 1.2×10^{-3}, which is already better than the QEF point for the acceptable reception. The bit error rate, thus, had been low enough for the correct reception of the signal. In this example, the amount of PE was between 7 and 75, and the averaged received power level was measured and averaged to -87 to $-89\,$dBm. It is worth noting that the RSSI resolution is 1 dB for single measurement event in the used version of the field test software.

Figure 10.30 shows the measured RSSI values during the complete test route. There were two rounds done during each test. The back lobe area of the test route can

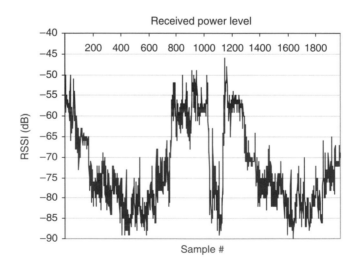

Figure 10.30   The RSSI values measured during the test route

be seen in the middle of the figure, with fast momentarily drop of received power level. The received power level was about -50 dBm close to the site, and about -90 dBm in the cell edge. The duration of the single test route was approximately 25 min, and the total length of the route was 22.4 km.

The maximum speed during the test route was about 90 km/h, and the average speed was measured to 50 km/h (excluding the full stop periods). The speed is sufficient for identifying the effect of the MPE-FEC.

10.6.1.4   Method for the Analysis

The collected data was processed accordingly in order to obtain the breaking points, i.e., the QEF of 2×10^{-4} and FER/MFER of 5% in function of the RSSI values for each test case. The processing was carried out by arranging the occurred events per RSSI value. For the BB and BA, the values were averaged per RSSI resolution of 1 dB. For the FER and MFER, the values represent the percentage of the erroneous frames compared to the total frame count per individual RSSI value (with the resolution of 1 dB).

Figure 10.31 shows an example of the processed data for the bit error rate before and after the Viterbi for 16-QAM, CR 2/3, MPE-FEC 2/3, and FFT 2K. The results represent the situation over the whole test route in location-independent way, i.e., the results show the collected and averaged BB and BA values that have occurred related to each RSSI value in varying radio conditions.

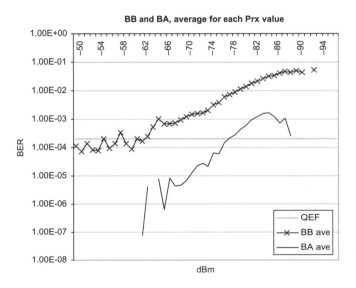

Figure 10.31   Postprocessed data for the bit error rate before and after the Viterbi presented in logarithmic scale

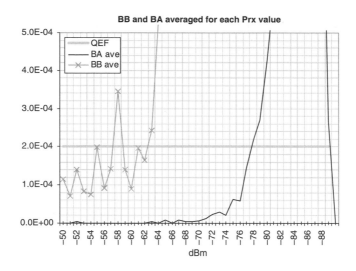

Figure 10.32 Processed data for the bit error rate before and after the Viterbi with an amplified view around the QEF point in linear scale

As can be noted in this specific example, the bit error rate before Viterbi does not comply with the QEF criteria of 2×10^{-4} even in relatively good radio conditions, whereas the Viterbi clearly enhances the performance. The resulting breaking point for the QEF with Viterbi is found to be approximately -78 dBm of RSSI in this specific case.

For the FER, the similar analysis yields an example that can be observed in Figure 10.32. This figure shows the occurred frame error counts (FER and MFER) as well as the amount of error-free events analyzed separately for each RSSI value. In this format, Figure 10.32 shows the amount of occurred samples as a function of RSSI in 1 dB raster arranged to error-free counts ("FER0, MFER0"), to counts that had error but could be corrected with MPE-FEC ("FER1, MFER0"), and to counts that were erroneous even after MPE-FEC ("FER1, MFER1").

It can be noted that the amount of the occurred events is relatively low in the best field strength cases and does not necessarily provide with sufficient statistical reliability in that range of RSSI values. Nevertheless, as the idea was to observe the performance especially in the limits of the coverage area, it is sufficient to collect reliable data around the critical RSSI value ranges.

In this type of analysis, the data begin to be statistically sufficiently reliable when several tens of occasions per RSSI value are obtained, preferably around 100 samples as stated in [Bou06]. In practice, though, the problem arises from the available time for the measurements; i.e., in order to collect about 100 samples per RSSI value in large scale, it might take more than 1 h to complete a single test case. There was a total 32 test drive rounds carried out, 25 min each. In this case, the

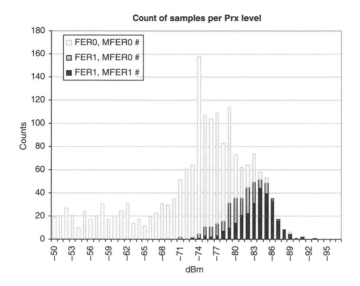

Figure 10.33   Example of the analyzed FER and MFER levels of the signal

postprocessing and analysis was limited to two terminals, though, due to the extensive amount of data.

The corresponding amount of total samples was normalized, i.e., scaled to 0–100% separately for each RSSI value. An example of this is shown in Figures 10.33 and 10.34.

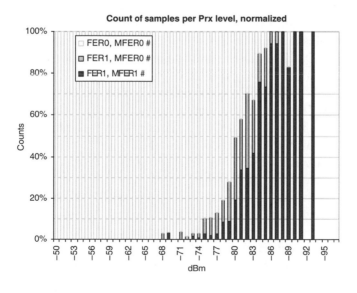

Figure 10.34   The postprocessed data can be presented in graphical format with FER and MFER percentages for each RSSI value

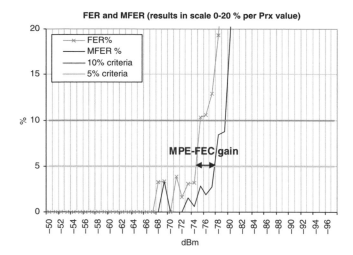

Figure 10.35   An amplified view to FER5 and MFER5 criteria shows the respective RSSI breaking points

By presenting the results in this way, the percentage of FER and MFER per RSSI and, thus, the breaking point of FER/MFER can be obtained graphically.

The 5% FER and MFER levels, i.e., FER5 and MFER5, can be obtained graphically for each case observing the breaking point for the respective curves (Figure 10.35). The corresponding MPE-FEC gain can be interpreted by investigating the difference between FER5 and MFER5 values (in dB). The graphics show the observation point directly along the 5% error line. The parameter values of the following examples are still 16-QAM, CR 2/3, MPE-FEC 2/3, and FFT 2K.

As stated in the DVB-H implementation Guidelines [Dvb06], the moving channel produces fast variations already in TU6 channel type (typical urban 6 km/h) in the QEF criterion making the correct interpretation of the bit error rate before Viterbi very challenging. This also leads to the uncertainty of the correct calculation of the bit error rate after Viterbi. Thus for the bit error rate before and after Viterbi, it is not necessarily clear how the terminal calculates the BB and BA values especially in the cell edge with high error rates, and the frame error analysis gives more reliable means of interpreting the quality.

The benefit of the hand-held receiver is obvious for the measurements as the equipment is easy to carry to different environments, including indoors. The data collection with hand-held terminal is fast, and the collected radio interface performance indicators provide sufficient data for the postprocessing. The terminals can be used as an additional tool for fast revision of the overall functioning of the network. With the collected data and respective postprocessing, it is possible to observe the DVB-H audio/video quality in detailed level compared to the subjective studies.

10.7 EMC and Bio-effect Calculations

10.7.1 Introduction

The radiation of the DVB-H transmitter sites can be assumed to be normally higher than in mobile networks. Depending on the antenna installation and gain, the radiating power can be more than 50 kW. As there are normally other systems installed in the same site as DVB-H, e.g., mobile or broadcast systems, it is essential to dimension the safety distances in order to minimize the risk of intersystem interferences. On the other hand, the human exposure limits must be calculated for both technical installation personnel as well as for the public.

The DVB-H transmitter site antennas can be located to the telecom or broadcast tower as well as on the rooftop of the buildings. Furthermore, the low-power and gap-filler sites can use indoor installations for the antennas. Depending on the environment, the power level should be adjusted accordingly in order to avoid any interferences or safety zone extensions. Section 10.7.2 presents the methodology to estimate the safety distances in typical environments.

10.7.2 Safety Aspects

The power level of DVB-H depends on the transmitter manufacturer models. Typically, the power amplifiers can produce around 100–9000 W (output power fed to feeder), although the maximum power level might be limited to few thousands of watts in practical solutions.

DVB-H radiation is nonionizing like in case of mobile network technologies. Thus the radiation does not alter the human cell structure as can happen in ionizing systems like X-ray equipment. Nevertheless, a sufficiently high nonionizing radio transmission power can increase the cell temperature.

The radiation in given distance can be estimated with various propagation models. The simplest one giving the maximum values is the far-field attenuation in free space:

$$L = 32.44 + 20\log(d) + 20\log(f) \qquad (10.1)$$

The safety distance of the DVB-H antennas must be assured in the deployment of the system. A simple but practical method can be created based on the fact that the antenna is located to the tower or rooftop, and that the most meaningful area to be investigated is found below the antenna.

The DVB-H antenna solution can be based on omni-radiating poles or directional antennas. Figure 10.36 shows an example of the far-field horizontal and vertical radiation patterns of the directional antenna used in DVB-H. As can be observed from the figure, in the practical installation environments, the vertical

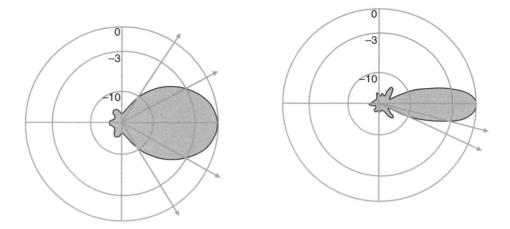

Figure 10.36 Example of the horizontal plane of the antenna radiation pattern. The 3 dB attenuation determinates the beam width. In this specific case, the beam width is about 60°. The second pattern shows an example of the vertical plane of the directional antenna radiation. In this case, the beam width in 3 dB attenuation points is about ±14°, i.e., 28°

radiation pattern is the most meaningful when estimating the field strength and safety zones.

In order to obtain the safety limits below the DVB-H antenna, a loss analysis with different angles from the antenna can be done. Having the vertical antenna pattern, respective coordinate system and linear scale for the antenna radiation attenuation as shown in Figure 10.37, an antenna attenuation table can be created with the scale

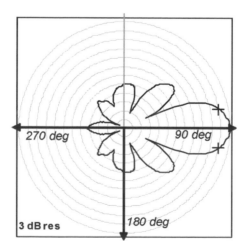

Figure 10.37   An example of the vertical pattern in linear scale with 3 dB resolution

Table 10.6 Example of the antenna attenuation values (in dB)

Degree	A_v	Degree	A_v	Degree	A_v
270	23.1	210	26.0	150	23.1
260	26.0	200	24.4	140	16.5
250	30.5	190	21.9	130	14.0
240	28.0	180	20.9	120	20.0
230	24.4	170	20.9	110	10.8
220	24.4	160	30.5	100	2.2

from 270° to 180° (back lobe) and from 180° to 90° (main beam). Here 180° vertical angle means the point below the antenna. In this specific example, Table 10.6 can be obtained from the respective antenna data. The 90° angle represents the main beam of the antenna with 0 dB loss. Please note the difference with the conventional marking of the angles, as normally 0° elevation represents the horizon and − 90° the point below the antenna.

The values of the table can now be analyzed in two phases. The first task is to obtain the minimum attenuation value for the back and side lobes, i.e., in beam angle of 270°−180°. In this example, it is 20.9 dB. Since the minimum attenuation of the horizontal pattern falls into this value, it can be used as a reference for the complete back side (hemisphere) of the antenna. The vertical angle between 180° and 90° is used for the analysis of the main lobe. The angle is the independent variable used to obtain the respective safety distances.

The next step is to calculate the safety distance for the radiation power, which is the result of the DVB-H transmitter power, transmitter filter, cable, connector and jumper losses, possible power splitter loss for multiple antenna arrays and the radiation pattern loss in function of the elevation angle. The safety distance depends on the regulatory decisions. As a common type of regulations, Equation (10.1) can be used in order to obtain the respective formulas.

The immediate area close to the antenna is the near-field region. Most of the electromagnetic energy in this region is stored instead of radiating. The field has considerable variations within this zone, making the field estimation extremely challenging. Further away from the antenna, the reactive near-field decreases and the radiating field becomes predominating as a function of the distance until the far-field zone finally stabilizes the characteristics of the radiation making the calculation of the field strength reliable.

The dimensions of the radiating antenna have impact on the minimum distance where the far-field starts dominating. Assuming the variable D indicates the largest dimension of the antenna and if λ is the wavelength of the observed signal, the following formula can be used for the calculation of the minimum far-field limits in case of large antennas (i.e., if the greatest dimension of the antenna is much more than the wavelength), according to Equation (10.1):

$$d_{\text{far-field}} = \frac{0.5D^2}{\lambda} \tag{10.2}$$

For small antennas, following formula can be used:

$$d_{\text{far-field}} = \frac{\lambda}{2\pi} \tag{10.3}$$

The next calculations are valid for the far-field, in frequency range of 300–1500 MHz, which falls into the operating frequency range of DVB-H. In this specific case, the maximum power density for the general public in the DVB-H frequency range can be obtained by the following formula:

$$W = \frac{f(\text{MHz})}{150} \tag{10.4}$$

Now, the power density in the far-field region can be obtained by the following formula:

$$W = \frac{\text{EIRP}}{4\pi r^2} = \frac{PG}{4\pi r^2} \tag{10.5}$$

In the formula, EIRP is the effective isotropic radiated power (in watts), r is the distance from the radiating antenna (in meters), P is the power fed to the antenna and G is the antenna gain compared to the isotropic antenna ($10\log_{10}(G)$ in dBi).

It is now possible to estimate the safety distance in front of the DVB-H antenna. Investigating Table 10.6, it is possible to estimate the safety distances also in different sides of the antenna. Let us select the 700 MHz frequency and transmitter power of 1500 W after the filter loss. The additional cable, connector, and jumper loss (L) can be estimated to be 3 dB. Using the antenna pattern presented in Figure 10.37, assuming its gain (G) to be 13.64 dBi and the physical dimensions to be $190 \times 500 \times 1000$ mm, the power density and EIRP values are as follows:

$$W = \frac{700 \text{ MHz}}{150} = 4.67 \text{ W/m}^2 \tag{10.6}$$

$$\text{EIRP}(W) = P \times 10^{\left(\frac{G-L}{10}\right)} = 1500 \text{ W} \times 10^{\left(\frac{13.64-3}{10}\right)} \tag{10.7}$$
$$\approx 17.38 \text{ kW}$$

The minimum safety distance in main beam is thus

$$r = \left[\frac{\text{EIRP}}{4\pi W}\right]^{0.5} = \left[\frac{17.38 \text{ kW}}{4\pi \cdot 4.67 \text{ W/m}^2}\right]^{0.5} \tag{10.8}$$
$$\approx 17.2 \text{ m}$$

The wavelength of the used frequency is

$$\lambda = \frac{c}{f} = \frac{300,000 \text{ km/s}}{700 \times 10^6 \text{ Hz}} = 0.43 \text{ m} \tag{10.9}$$

where c is the speed of light.

Let us revise the minimum distance for the far-field in order to make sure that the calculation is valid. As the extreme dimension of the antenna is more than the wavelength (1 m), the antenna is considered large, and the far-field distance limit is thus

$$d_{\text{far-field}} = \frac{0.5D^2}{\lambda} = \frac{0.5 \cdot (1 \text{ m})^2}{0.43 \text{ m}} \approx 1.2 \text{ m} \tag{10.10}$$

The calculation is valid as the value is in far-field zone. For the radiation angles of $180° - 90°$, Table 10.7 can be created.

For the vertical back and side lobes, i.e., for the vertical radiation angles of $\alpha = 270°-180°$, the common value of 20.9 dB was obtained from Table 10.6. This value corresponds to the minimum safety distance of 1.6 m for the whole range of the above-mentioned angles. As can be observed from Tables 10.6 and 10.7, the relatively narrow vertical beam width provides reduced safety distances outside the main lobe. Figure 10.38 clarifies the idea of the safety distances in graphical format.

When using a 2-m high antenna pole, the safety distance below the antenna is thus achieved for the personnel inside the building. As the antenna is in the edge of the rooftop, the main beam is also secured. In addition, the roof material provides with additional attenuation for the indoor.

Figure 10.39 shows and extended analysis with the same radio parameters as previously but varying the transmitter power level between 100 and 9000 W.

As can be observed, lower the power level and closer to the back side of the antenna, many calculated points will fall below the minimum far-field distance of 1.2 m. As the calculation is not accurate in those near-field spots, the 1.2-m limit can be considered as a minimum limit for all the cases in near-field.

In case of Figure 10.40, the safety limits on the rooftop should be marked accordingly in case of, e.g., the maintenance personnel closing the antenna. The back side of the antenna can be marked as round with the minimum calculated safety distance, with preferably extra margin as the antenna pole can affect the final

Table 10.7 Example of the safety distance values for the angles $180° - 90°$ with the respective vertical pattern attenuation

Degree	A_v	r_{min}	Degree	A_v	r_{min}
180	20.9	1.6	130	14.0	3.4
170	20.9	1.6	120	20.0	1.7
160	30.5	0.5	110	10.8	5
150	23.1	1.2	100	2.2	13.4
140	16.5	2.6	90	0.0	17.2

radiation pattern of the antenna (i.e., the pole might act as a part of the antenna elements).

For the main beam, the horizontal radiation pattern must be taken into account by calculating the safety limits on the sides of the antenna. A mask with sufficient additional margin can be used, e.g., by observing the reference attenuation points of 20, 10, and 3 dB and the respective angles. In practice, it is important to assure that the personnel cannot cross the main beam accidentally as the full EIRP might mean considerable safety distances in front of the beam.

10.7.3 EMC Limits

When the DVB-H antenna is installed in the tower, there might be antennas of the other telecommunication systems on top and below of the DVB-H antenna. Thus the

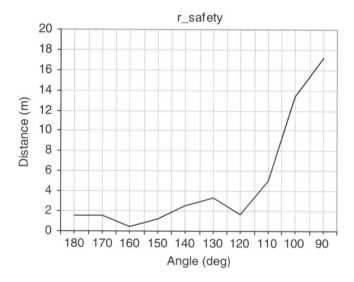

Figure 10.38 An example of the safety distances as a function of the angle of the vertical antenna radiation pattern

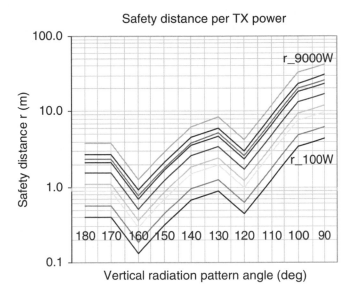

Figure 10.39   The safety distances for a set of transmitter power levels presented in logarithmic scale, with the parameter setting of the presented example

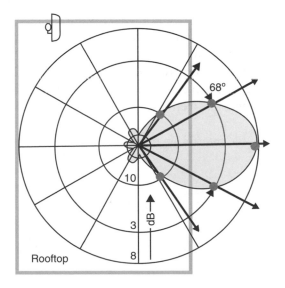

Figure 10.40   In case of the rooftop installation, the respective safety distance on sides of the antenna should be calculated according to the horizontal radiation pattern of the antenna

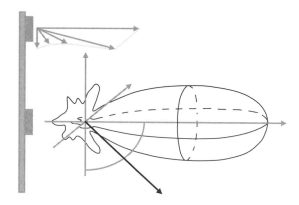

Figure 10.41 When using a directional antenna, the relative field strength above and below the antenna depends on the attenuation of the vertical radiation pattern

EMC calculations should take into account mainly the upper and lower side lobes of the DVB-H antenna.

The analysis methodology presented in Section 10.6 can be also used for the EMC calculations. Figure 10.41 shows a principle of the EMC in the towers. The DVB-H antennas create a certain electromagnetic field strength around the antenna, which might be considered as an interference field for the antennas of the other systems nearby.

Like in the previous case, the vertical radiation pattern can be used as a basis for the EMC field estimations. In fact, the vertical pattern right above and below the antenna is the most important as the antennas of the other systems are normally in the same vertical line of the tower.

As can be observed from Figure 10.42, the vertical pattern in this case is symmetrical above and below the antenna element. The attenuation value around 180° (below the antenna) as well as 0° (above) is thus 20.9 dB as shown in Table 10.1.

In order to calculate the interfering electrical field, the following formula can be used:

$$E_i = \frac{1}{2r}\sqrt{\frac{\eta P}{\pi}} \tag{10.11}$$

The variable η is the wave impedance in air with the value of 377 Ω, P is the transmitter power (EIRP) including the gains G and attenuations A.

Let us revise the electrical field for the previously presented case examples with the transmitter power levels of 1500 W in 2 m distance of the main beam. As shown previously, the example yields a total EIRP of 17.83 kW, which produces the

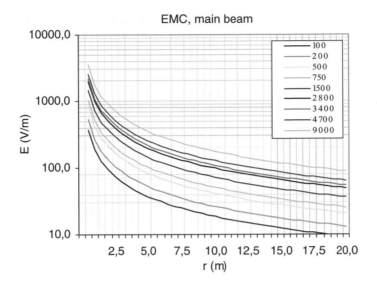

Figure 10.42 The electrical interference field in the main beam of DVB-H antenna with the transmitter power levels of 100–9000 W (corresponding EIRP of 1.2–104 kW)

following field in 2 m distance in the main beam:

$$E_i = \frac{1}{2r}\sqrt{\frac{\eta G P}{\pi}} = \frac{1}{4}\sqrt{\frac{377 \times 17830\,\text{W}}{\pi}}$$

(10.12)

$$\approx 366\,\text{V/m}$$

The field strength reduces according to the radio propagation environment. Assuming the worst case, the free propagation loss can be assumed as shown in Equation (10.1).

The antennas of the other systems might be relatively near the DVB-H antennas due to the lack of the space in the tower. The essential question is thus what is the minimum allowed distance between the antennas.

The allowed field strength depends on the immunity level of each system. As an example, the commercial electrical equipment are able to handle interference field strengths of about 2–10 V/m. For the professional electrical equipment as well as, e.g., mobile network elements, the requirement is considerably higher.

In addition to the pure field strength, the frequency is also essential. As an example, the GSM 850 and GSM 900 might suffer from the distortion caused by the harmonic components of the DVB-H transmission in the radio spectrum, whilst GSM 1800/1900 is safe due to the frequency difference and filtering of the equipment.

For the situation with the antennas located on top of each others, the attenuation value of the vertical radiation pattern of the DVB-H should be taken into account. The value for the EMC calculations can be estimated by observing the 180° angle of the vertical radiation pattern. In practice, the value might need additional margin in order to take into account the possible variations, e.g., for the pattern inequality around 180°, and the possible side effect of the tower itself, which can interfere the radiation pattern. The effect of the radiation pattern can now be extended in the following way.

$$P_{tot}(\mathrm{dBm}) = P(\mathrm{dBm}) + G(\mathrm{dB}) - A_v(\mathrm{dB}) \qquad (10.13)$$

$$P_{tot}(W) = \frac{10^{\frac{P_{tot}(\mathrm{dBm})}{10}}}{1000} \qquad (10.14)$$

$$E_i = \frac{1}{2r}\sqrt{\frac{\eta P_{tot}(W)}{\pi}} \qquad (10.15)$$

The A_v represents the attenuation (dB) in the observed vertical direction; in this case, the value is 20.9 dB both right above as well as below the antenna, as can be seen in Table 10.1.

The severity of the interfering field depends on the frequency, i.e., how considerable the frequency separation is.

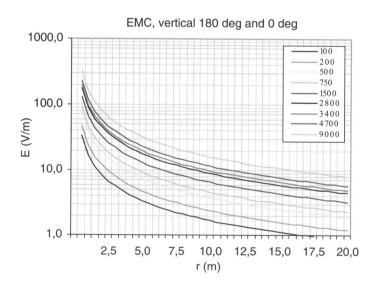

Figure 10.43 The electrical interference field above and below the typical directional DVB-H antenna with the transmitter power levels of 100–9000 W (corresponding EIRP of 9.4–848 W)

The effect of the main beam should be observed especially when the DVB-H antenna faces the antenna of another system. The EMC safety distance is sufficient in most of the installation cases as there should be sufficiently open area in front of the antenna, but there might be situations, e.g., with antennas located on rooftops in both sides of a street. It is thus important to make sure that the facing antennas (e.g., DVB-H antenna on rooftop and GSM 850 antenna on the other side) are producing sufficiently low field within the respective distance. The field in the main beam might be quite high as can be observed from Figure 10.43; meaning that the mid-level and the highest power transmitters should be avoided in rooftops.

11

Optimization of the Network

The complete DVB-H system consists of DVB-H radio network and respective core network that can be referred as a combination of head end and respective transport network, or simply as DVB-IPDC in the specific case of IPDC mode. The latter also include system and element management methodologies. The end-to-end solution delivers the data stream from the encoder (that captures it from the source element) up to terminals.

In order to design an optimal network, it is thus important to balance all the parts, i.e. elements and interfaces of the solution accordingly, taking into account the effects of the functionality, possibilities and limitations. In addition, the inter-dependencies of the parameter settings should be known. An optimal setting of one part of the network might result in low-quality performance in other parts of the network. As an example, a too high bit rate in encoder's output limits the total channel capacity, and can even lower the quality of the encoded stream.

The in-depth technical network planning also includes the economical optimization of the network. The effects can be investigated e.g. in function of the cost and transmitter power levels.

In addition to the good balancing of the coverage and capacity, the optimization also should include the usability considerations. One example of the functionality affecting the user perception is the channel change time, which depends on the time slicing setting. As in the large roll-out of network sites the total cost is considerable, it is important to understand what is the most cost-efficient way of deploying the services. The balance between technical quality and costs requires information regarding where the potential users could be found.

The variables for the complete techno-economic network optimization include the total cost of the site, radius of the cell and radiating power level vs. the respective

The DVB-H Handbook   Jyrki T.J. Penttinen, Erkki Aaltonen, Jani Väre and Petri Jolma
© 2009 John Wiley & Sons, Ltd

cost of the equipment. There are various sources of information as well as tools in order to identify the main aspects in the physical environment, including the geographical distribution of the population and economical levels, databases of the site locations (existing towers, the height of the antennas), digital maps with the area types and respective propagation models, etc. In the detailed network optimization, unlike in the nominal plan, uniform power levels would not provide the best performance due to the practical differences of the areas and site solutions, so the power levels and antenna heights should be planned site-by-site basis in this phase.

Technically, the higher the transmitter power and radiating power level, the better the DVB-H network. Maximizing the transmitter power, antenna gain, antenna height, and minimizing the losses, the cell radius is maximal, but on the other hand, high-power transmitter costs more and consumes more energy than lower ones, and greater height towers are more difficult to get than lower ones. The cost-efficient optimal point might be thus found somewhere between the extreme values of the final radiating power.

The maximum power level per site depends on the transmitter equipment and the respective antenna type. The transmitter equipment might be modular with variable amount of power amplifier elements. It is useful for the extensions of the network e.g. when QPSK modulation is changed to 16-QAM that requires higher transmitting power or higher antenna locations, or more sites in order to achieve the similar coverage areas. The site optimization might require even information about the number of the racks per transmitter as it has effects on the physical footprint of the equipment, and thus on the required floor space. Also the power consumption of each transmitter should be known which affects the OPEX and the required air cooling of the site.

As the ideal optimization should be balanced between the technical solutions and their costs, the following main site items should be known:

- antenna type vs. performance;
- antenna cable type vs. loss;
- transmitter power level vs. radius;
- filter loss (e.g. 10% can be used as a general rule).

The CAPEX and OPEX of the network depend, among the other items, on the transmission solutions between the main core network sites and the radio transmitter sites. In case of leased lines, the operating cost between the network elements is obviously higher than the one generated by own existing infrastructure. In the case of satellite transmission solution e.g. between the IP encapsulator and DVB-H transmitter, some optimised methods have been developed including the delivery of local content without wasting capacity.

In a large DVB-H network with multiple radio sites and without existing terrestrial transmission infrastructure, the most cost-effective transmission technique is probably based on the satellite links in short- and mid-term focus. The selection between satellite and terrestrial transmission depends obviously on the cost of each case in short and long term, taking into account the initial and future network deployment, i.e. the number of the sites and the distances between the sites.

The radio network roll-out costs depend heavily on the amount of new physical sites. It is thus obvious that more the already existing telecommunications and broadcast sites are re-used, lower the CAPEX and OPEX of the DVB-H network would be.

In the optimal functioning of the DVB-H network, the major question is relating to the correct selection of the site locations. Depending on the operator's other activities, there might be possibilities to reuse e.g. the existing mobile communications or DVB-T sites if there is only room for the extra DVB-H equipment. The antenna mounting could be done using the existing telecom or broadcast towers, although the optimal antenna height is probably not possible to obtain in major part of the cases. As an example, it is highly unlikely that the topmost part of the already existing telecommunications or broadcast towers would be available. One of the basic limitations in these cases is the electromagnetic compatibility (EMC), so the minimum distance between the DVB-H and other antennas should be calculated case-by-case in order to avoid any interference between different systems.

In the detailed planning of the network, it is thus important to investigate in advance which sites can be possibly used for DVB-H, and which are the practical limitations of the sites as the transmitter power levels, antenna types and antenna heights are considered. Other limitations could include the availability of transmission lines (either optical fibre, radio transmission or satellite links), and the availability of power especially in rural areas.

As for the terminals, only little can be done to optimize the performance. Normally neither the end user nor the operator can do much in order to optimise the coverage area with the terminal equipment tuning. Nevertheless, one of the most logical ways to increase the performance e.g. in the edge of the coverage areas is to use external antenna. The in-build antenna of the terminal is normally the limiting factor in DVB-H link budget. The rough estimation for the antenna gain might be in the order of -7 to -9 dBi. As the external directional antenna could provide normally a gain of $+5$ to $+10$ dBi, the increase in the performance is obvious if the mobile is situated e.g. in summer cottage type of relatively fixed position. Unfortunately this method is not very applicable in normal pedestrian use cases of DVB-H as the terminal is meant for the moving environment. Nevertheless, the benefits of external antenna are obvious in vehicle mounted equipment.

The optimization of the DVB-H network includes the selection of the most logical parameter settings for the core and radio subsystems. The tuning of the network might last the whole life cycle of the network due to the network and service set expansions and smaller fine tunings. As the DVB-H is a broadcast system, it is important to carry out sufficiently periodical field measurements to assure the continuous quality of service areas.

In the detailed DVB-H network planning and optimization, it is not sufficient to select technically best options. As in the normal case DVB-H requires its own infrastructure, the initial costs are relatively high. The cost is logically lower if already existing transmission infrastructure (transmission, sites, towers) of e.g. parallel DVB-T system can be utilized.

As an additional example, more the output power provided by the transmitter, higher the equipment complexity and power consumption, which affects the operating expenses of the network. Similarly, even technically more efficient, higher antenna location in broadcast tower might be more expensive than with the antenna placed on rooftop. In the optimal deployment of the network, it is thus essential to identify the most relevant technical parameters and investigate their impacts on the initial, i.e. CAPEX and OPEX. Even in case of relatively small DVB-H network deployment, the proper election of the transmitter types (power levels) might reduce considerably the final costs of the network both in short and long term, making the operation of the network more profitable (Figure 11.1).

The following sections describe the DVB-H optimization via different technical parameter strategies based on e.g. the error coding, single frequency network, transmitter powers and antenna heights.

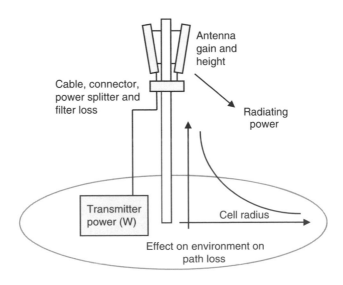

Figure 11.1 The main aspects of the transmitter site deployment that should be taken into account in DVB-H network's link budget planning

11.1   Technical Parameters

11.1.1   Modulation

The modulation scheme can be varied depending on the capacity and coverage requirements. Within a single cell area (a group of physical site cells in SFN area), the modulation should be logically kept the same, but as one example, an isolated area with a single cell that works in different frequency, i.e. in a separate MFN area where the entrance is done via handover, could use different modulation scheme than the main network.

One of the cases that could be realized in practice is the variation of the modulation in separate frequencies e.g. in such a way that 64-QAM would be in use inside an isolated shopping centre whilst the rest of the area uses either 16-QAM or QPSK. Although 64-QAM is very sensible for the variations in radio interfaces like impulse noise and Doppler shift, the area within the shopping centre is adequate for the modulation as it provides sufficiently controlled and static or slowly moving mobiles in the pedestrian environment.

In the normal DVB-H constellation case, mapping of the data onto OFDM symbols refers to the individual modulation of each sub-carrier based on one of the three basic DVB-T modulation schemes: QPSK (which is also referred as 4-QAM), 16-QAM or 64-QAM. Depending on the constellation chosen, 2 bits for QPSK, 4 bits for 16-QAM and 6 bits for 64-QAM are carried at a time on each sub-carrier. It can be estimated that QPSK is about four to five times more tolerant to noise, than 64-QAM.

The hierarchical modulation could be a usable strategy in some cases. It basically uses the sub-set of higher modulation scheme, and thus provides more flexibility to the selection of the modulation schemes. Hierarchical modulation constitutes an alternative usage of the 16-QAM and 64-QAM constellations. It can be said that hierarchical modulation consists of two separate 'channels' that are separated in RF level, and each one of these channels consists of different capacity and robustness (C/N requirement). This means that the two channels form the coverage areas that differ form each others. In practice, one of the data streams is mapped by using QPSK constellation. The bit pairs of the respective data stream define the quadrant of the sub-carrier within the constellation. The other data stream is used to modify the defined quadrant, i.e. the real and imaginary components of the sub-carrier. If the second data stream is mapped by bit pairs, the hierarchical constellation is called 'QPSK over QPSK' and the resulting constellation looks like a 16-QAM. If instead 4 bits are used, then it is called '16-QAM over QPSK', and the result looks like 64-QAM constellation. Figure 11.2 clarifies the idea of the hierarchical modulation.

In the hierarchical modulation, the first data stream always uses QPSK modulation. As it is the most robust, it is referred as a high-priority stream (HP). The second data stream, which is either QPSK or 16-QAM, is called as low-priority stream (LP) as it is less robust.

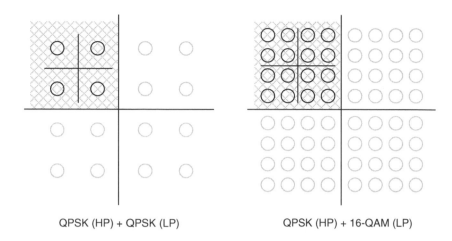

QPSK (HP) + QPSK (LP)   QPSK (HP) + 16-QAM (LP)

Figure 11.2 The basic idea of hierarchical modulation. The grey area indicates the high-priority stream (one out of the four possible areas), and the constellation within that area indicates the coding of the low-priority stream

The hierarchical modulation also provides an additional alpha factor. It is a spatial offset for each quarter of the constellation within the quadrant. The offset enhances the robustness of the QPSK/HP modulation, although the performance of the remaining LP modulation gets weaker.

The benefit of hierarchical modulation is that it allows the broadcasting of two independent transport streams on the same radio channel, and it provides with dedicated protection and respective dedicated coverage to each transport stream.

11.1.2 Balancing of Code Rate and MPE-FEC

One of the essential questions in the radio network planning and optimization is the balance of the code rate (CR) and MPE-FEC rate. It can be noted that the CR has a major impact on the quality of the DVB-H. On the other hand, the task is to decide how much capacity is offered, as it is directly proportional with the CR.

The main function of the MPE-FEC is to 'clean' as much as possible the remaining frame errors after the basic convolutional coding. The performance of MPE-FEC has been studied in different environments and with different setup of the radio parameters. As a rule of thumb, heavier the MPE-FEC rate, better the performance. The optimal MPE-FEC rate is not probably the most protected one though it uses largest amount of capacity, which reduces accordingly the useful bit rate.

The MPE-FEC performance can be carried out by post-processing and analysing of the radio measurement results. As the usefulness of the MPE-FEC depends on the environment, it might be interesting to carry out local performance studies even if the general guidelines of the MPE-FEC performance would be known. For this purpose, a respective field test method is presented in this chapter. The field test examples were carried out in different area types of Helsinki, Finland, inside partially overlapping coverage areas of a commercial DVB-H network as described in [Pen09b]. Also selected laboratory measurements were carried out in indoor environment. For the sub-urban environment, a respective investigation result set of USA is presented based on the [Pen08a].

In the quality revision of the network, the frame error rate (FER) indicates sufficiently well the real performance of the radio network as it is comparable with the end-user's interpretation of the service quality. More specifically, the DVB-H planning guideline [Dvb06] proposes that FER of 5% or less indicates that the reception is sufficiently good as interpreted by the users. The 5% limit has been selected based on a subjective estimation by assuming that single destroyed frame with a length of 1 s, during a total of 20-s interval, is still providing an acceptable quality level in practice. Furthermore, [Dvb06] shows that the 10% frame error level is already annoying in practical reception. After the first phase DVB-H evaluation tasks, more in-depth criteria have been searched for. As an example, [Him06] has identified that the limit for the acceptable quality based on the subjective interpretation is roughly between 6.9% and 13.8% as for the MFER when the channel type is close to vehicular urban (VU). This indicates that the MFER5 might be slightly pessimistic, but it is useful for the comparison purposes for the results obtained in scientific studies.

11.1.2.1   MPE-FEC in Indoor Environment (Laboratory Studies)

The level of the MPE-FEC correction can be obtained by investigating what is the level of frame errors (FER) compared to the frame errors after the MPE-FEC functionality (MFER, MPE-FEC FER).

Normally, the field measurement equipment does not include more in-depth ways of investigating the FER distributions in function of selectable variables. Nevertheless, if the FER can be stored in time line with the respective information of e.g. received power level, the post-processing can be carried out. In the sufficiently detailed analysis, the field test results showing FER before and after the MPE-FEC (MFER) are quite sufficient in order to obtain the RSSI limit of the used error correction scheme in that specific environment and case. As an example, the raw data can be collected by using logging; the list of test results is organized related to elapsed time, i.e. the information of the frame error or successfully received frame (e.g. with the indicator 0 showing no errors, and 1 showing that the current frame was erroneous) is shown in the log file as such for each consecutive frame.

This list of the field test results can be organized further by post-processing the log file into a PDF format, received power level representing the x-axis. As an example, the field test software of Nokia presents the RSSI measurement results

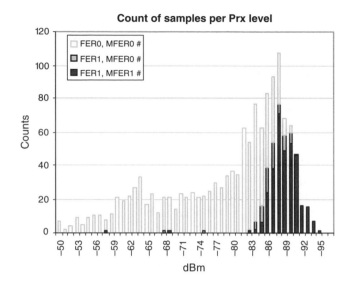

Figure 11.3 An example of the collected and post-processed data. The RSSI raster is 1 dB, and each RSSI value corresponds to the error-free samples (FER 0), as well as occurred frame errors (FER 1) that could be corrected by MPE-FEC (0) or not (1)

with 1 dB resolution, and respective FER and MPE-FEC results can be used as variables. The first post-processing round could result a table of occurred FER and MFER events per RSSI value as seen in Figure 11.3. When these values per received power level (RSSI) are normalized to 100%, the format shows the proportion of the occurred frame errors that could be corrected, arranged as functions of the RSSI levels as seen in Figure 11.4. The format gives indication about the general probability distribution of the MPE-FEC functionality in investigated area type in function of RSSI. For the comparison of the performance with the terminal speed, also a cumulative presentation was calculated for frame error occasions showing the expected success rates of MPE-FEC with different parameter values. This gives indication of the effectiveness of MPE-FEC per case in different area types, i.e. with what RSSI value MPE-FEC starts effecting, and up to what RSSI limit it is still able to correct sufficiently effectively the frame errors. An example of this format can be seen in Figure 11.5.

In case of Figures 11.3 and 11.4, the term 'FER1, MFER0' means that there has been frame error that could be recovered by MPE-FEC. 'FER1, MFER1' means that the occurred error could not be recovered any more, and 'FER0, MFER0' means that there were no frame errors present in the first place.

As an example of the usefulness of the methodology, Table 11.1 summarises the laboratory test cases presented in [Pen08e] and respective results for the MPE-FEC breaking points (i.e. showing the RSSI value with the 5% criteria) in received power levels (dBm) when using 16-QAM, CR of $^1/_2$, FFT of 8K and guard interval

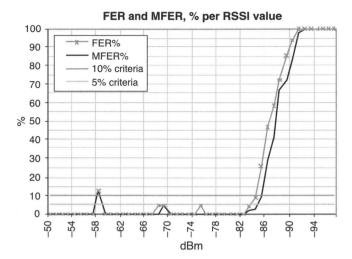

Figure 11.4 An example of the MPE-FEC analysis showing the occurred frame errors related to received power level, scaled to 0–100% of frame errors per RSSI value. In this format, each RSSI value (in 1 dB scale) is interpreted separately by normalizing the percentage of occurred frame errors

(GI) of 1/8. It should be noted that these results are merely snap-shot values for that specific case, and should be revised separately for the different environments where in-depth understanding about the variations of the MPE gain effect should be fine-tuned to the respective link budget.

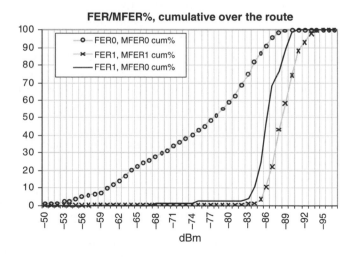

Figure 11.5 An example of the cumulative analysis of the measurement data. This format gives a rough indication about the RSSI scale where the MPE-FEC was taken place and where it was effective. As an example of this figure, the signal level of about − 83 to − 90 dBm produces frame errors that can be corrected. The scale of − 87 onwards already produces errors that cannot be corrected with MPE-FEC

Table 11.1 The DVB-H laboratory measurement results

Case/FEC (%)	Terminal	FER5	MFER5	Diff
A/15	1	− 78.0	− 78.05	0.05
B/25	1	− 77.1	− 77.5	0.4
C/35	1	− 68.8	− 70.7	1.9
A/15	2	− 83.2	− 83.4	0.2
B/25	2	− 81.0	− 81.0/−83.2	0.0/2.1*
C/35%	2	− 82.5	− 84.5	2.0

Note: The MPE-FEC gain was not possible to obtain explicitly due to the instability of the results in MFER5 point.

In this specific case study, there were two terminals used in the data collection, Nokia models N-77 (terminal 1) and N-96 (terminal 2), with a special channel display application. As the results show, the terminals were not commonly calibrated as for the RSSI display. Nevertheless, this is not important in this type of revisions as the relative frame error information with and without MPE-FEC per terminal can be obtained even without the exact RSSI breaking point.

As Table 11.2 shows, the difference between FER5 and MFER5 increases as the MPE-FEC rate grows, i.e. there is a logical behaviour noted for the frame error capability in function of the MPE-FEC rate in all the cases. In the least protected mode (MPE-FEC 15%), the difference, i.e. the MPE-FEC gain, is close to zero. This can be explained by the building layout, which did not allow too much multi-path components to occur in the receiving end. In the most protected mode that was applied in these measurements (MPE-FEC 35%), the MPE-FEC gain is in the order of 2 dB. The MPE-FEC rate of 25% seems to result in only approximately 0.5 dB gain.

This case study shows that the MPE-FEC gain is not significant in indoor environment with low amount of reflected multi-path components in radio interface and with slowly moving terminal, if the MPE-FEC rate is defined to 15% or 25%. In this specific study, the effect starts to be notable when using MPE-FEC rate of 35%, giving about 2 dB gain compared to the situation without MPE-FEC in the investigated channel types (which are close to typical urban 3, or TU3 type of models). On the other hand, 35% MPE-FEC rate reduces accordingly the data channel capacity.

Table 11.2 The DVB-H laboratory measurement results with a single terminal (the difference of the FER 5% and MFER 5% levels (in dBmscale of RSSI) indicates the MPE-FEC gain)

Case/FEC	FER5	MFER5	Diff
15%	− 78.0	− 78.05	0.05
25%	− 77.1	− 77.5	0.4
35%	− 68.8	− 70.7	1.9

Table 11.3 The DVB-H indoor measurement results

Case	Counts	Max speed (m/s)	FER5	MFER5	Diff
1	1022	1–2	− 84.5	− 85.5	1.0
2	556	1–2	− 79.0	− 80.1	1.1
3	518	1–2	− 87.4	− 88.0	0.6
4	506	1–2	− 85.4	− 85.4	0.0
5	506	1–2	− 87.1	− 87.3	0.2
6	515	1–2	− 88.3	− 88.8	0.5

In order to compare the publicly available material and test results obtained via different methodologies, some of the logical sources of information are found e.g. in IEEE Explorer, DVB-H Project office and CELTIC WingTV documentation.

11.1.2.2 Indoor Tests in Live Network

The presented methodology to analyse the MPE-FEC gain can be applied also in live network. Based on [Pen08e], Table 11.3 summarises some of the post-processed indoor measurement results for the city area of Helsinki, i.e. the breaking points of FER 5% and MFER 5% in function of the received power level (dBm) when the MPE-FEC rate has been fixed to 15%, i.e. the lightest MPE-FEC rate was used. Due to the nature of the network type that was already operative, this value was not possible to change. The ideal situation would be to investigate the effect by varying the MPE-FEC rate and by observing its effects on the performance.

In case of any other field test cases and respective analysis, the area can be divided into the most common types based on the building size, openness of the building, etc. The cases presented in [Pen08e] are based on the following practical building type division:

'1' represents a ground floor of a shopping centre with relatively large open areas. It can be assumed that some multi-path propagated components arrived to the measurement routes.
'2' is the ground floor of a very large shopping centre in the middle of the city with partially limited areas for the multi-path propagated components.
'3' is the upper floor of the case '2'.
'4', '5' and '6' represent the ground, second and third floor, respectively, of university campus building in sub-urban area just outside the city.

The environment also contained some occasional impulse noise sources that were caused probably by power systems of the mechanical escalators. Figure 11.6 presents the behaviour of the effect, and shows the performance of MPE-FEC in this type of situations. Even if the MPE-FEC gain is not possible to analyse

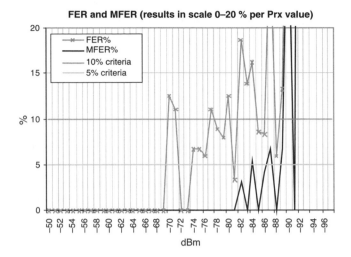

Figure 11.6 An example of the impulse type of noise that affected the measurement analysis in case B (note 1). In this case, the MPE-FEC was able to correct practically all the occurred frame errors up to about − 89 dBm RSSI level when MFER5 criteria is applied, but the actual MPE-FEC gain is not possible to interpret explicitly

explicitly by applying the methodology described in this chapter, this specific case shows the high usefulness of MPE-FEC when impulse noise is present if only the duration of the interference is within the functional limits of MPE-FEC.

When applying this investigation method, it should be noted that the amount of collected data must be big enough. If the number of samples is low, the statistical inaccuracy grows accordingly. The accuracy is statistically sufficiently high if each one of the observed RSSI values gets several tens, near 100 samples within that RSSI range where the error level grows exceeding the 5% error rate. It should also be noted that if the field strength is good, the MPE-FEC functionality does not 'wake up' and the breaking point of MPE-FEC cannot thus be observed.

As a general note of these cases, the results obtained from the laboratory environment are lightly more pessimistic compared to the dense urban indoor results. With 15% of MPE-FEC rate and common parameter set, the laboratory showed 0.2 dB gains whereas city centre's indoor environment provided normally up to 1 dB gains. This can be explained by the varying multi-path components in practical radio interface, the laboratory being more limited in this sense. In the sub-urban environment, though, the indoor gain was around in the same low level as was obtained in laboratory because there was only one main beam of the transmitter found in the area.

By applying this methodology, it can be expected that there are similar type of variations also in other cases which means that in detailed network planning and optimization, it would be justified to carry out this type of measurements in some of the most typical environments. The terminal speed can be varied from static to highest practical velocities. The resulting MPE-FEC gain can then be tuned accordingly to the local link budget.

11.1.2.3   Vehicular Tests

The results of e.g. [Pen08e] show that the normal terminal speeds in vehicular environment are not limiting the Doppler shift unless the most sensitive modes are in use. The theoretical speed limits and related parameter settings can be seen in Chapter 9, although in practice, even the 8K mode is normally sufficient for the typical vehicular usage.

11.1.2.4   Recommendations

As the MPE-FEC reserves a direct proportion of capacity from the total radio channel bandwidth, the balancing of the capacity and the MPE-FEC gain should be planned accordingly in the operational environment. The example presented in this chapter shows one possibility to measure and analyse the effect of MPE-FEC. The investigation of the MPE-FEC gain in real life is obviously the most reliable when wanted test cases with all the possible MPE-FEC rates can be carried out, combined with the most logical CR values. It is thus recommendable to perform respective field tests in the most typical environments where the potential users are found, in order to obtain the MPE-FEC gain as a basis for the respective link budget.

The laboratory test cases showed in this chapter indicates the behaviour of the MPE-FEC in one kind of indoor case, showing that the MPE-FEC functions even with the lowest MPE-FEC rates in the critical field strength areas although with only a small gain. On the other hand, the field tends to drop fast in the cell edge areas of indoor environment due to the propagation loss behind each corner, so MPE-FEC would probably not bring too much additional coverage as such in those indoor spots. Nevertheless, there exists sometimes impulse type of noise, which seems to be possible to correct even with low MPE-FEC rates as seen in Figure 11.7. It is thus an important decision of the operator to select the MPE-FEC rate correctly, not wasting too much capacity (e.g. by selecting the most efficient MPE-FEC rate of 50%) but assuring that the errors are corrected sufficiently (by investigating what are the benefits of other MPE-FEC rates compared to the capacity loss).

As a conclusion, MPE-FEC improves mobile DVB-H receiver performance as for the C/N and maximum achievable Doppler. The laboratory and indoor live network test cases presented in this chapter as well as in other publicly available documentation show that the Rayleigh channel effects as well as the impulse type of noise can be reduced with MPE-FEC. It is thus important to carry out the test

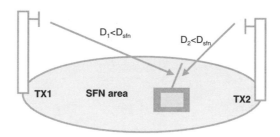

Figure 11.7 If all the sites of certain frequency band are located inside the SFN area with the distance between the extreme sites less than D_{SFN}, no inter-symbol interferences are produced

cases taking into account the maximum probable terminal speed in the planned area and with the selected parameter combination. As an example, according to [Dvb06], the 'typical reference receiver' should function up to 85 km/h speeds when using 16-QAM, CR $^1/_2$ and FFT of 8K, and 'possible reference receiver' should function up to 120 km/h speeds when using these parameter values.

11.1.3   Time Slicing

The tuning of the time slicing does not have direct impact on the radio performance, but the main issue is related to the usability of the service as the length of the time slicing affects the time the user has to wait until the channel change takes place. There is a balance for the sub-channel, i.e. time slice that should be obtained via the detailed network planning.

Even the larger time slice periods save accordingly more battery life per usage time of the DVB-H receiver, it is important that the battery optimization is not prioritized over the channel switching time. The users might compare the mobile TV service with the fixed TV experiences, which means that the waiting period in channel switching cannot be too long. It has been noted that a channel change time exceeding about 2 s is considered to be annoying. In practice, the amount of sub-channels could thus be e.g. 1 for the TV type of streams, and perhaps 2–3 for the audio (radio) services. The feasible values can be investigated e.g. in the pre-commercial phase of the network trials and pilots by collecting subjective opinions directly from the users in order to avoid the negative perception of the too long waiting periods in channel switching.

11.2   SFN Size

The DVB-H coverage area depends mainly on the area type, i.e. on the radio path attenuation, as well as on the transmitter power level, antenna height and radio parameters. The latter set has also effect on the audio/video capacity. In the detailed network planning, not only the coverage itself is important but the quality of service level should be dimensioned accordingly.

This section describes the SFN gain related items as a part of the detailed radio DVB-H network planning. The emphasis is put to the effect of DVB-H parameter settings on the error levels caused by the over-sized SFN area. In this case, part of the transmitting sites converts to interfering sources if the safety distance margin of the radio path is exceeded. A respective method is presented for the estimation of the SFN interference levels. The functionality of the method was tested by programming a simulator and analysing the variations of carrier per interference distribution. The results show that the theoretical SFN limits can be exceeded e.g. by selecting the antenna height in optimal way and accepting certain increase in the error level. Furthermore, by selecting the relevant parameters in correct way, it can be showed that the balance between SFN gain and interference level can be planned in controlled way.

As DVB-H is meant for the mobile environment, the respective terminals are often used on a street level for the reception. This creates a significant difference in the received power level compared to DVB-T, which uses fixed and directional rooftop antenna types. Furthermore, the DVB-H terminal has normally only small, in-built panel antenna, which is challenging for the reception of the radio signals. In the case of DVB-H, the factors affecting the quality are mainly related to the radio interface due to the variation in the received signal levels as well as the Doppler shift, i.e. the speed of the terminal.

The DVB-H service can be designed using either SFN or MFN modes. In the former case, the transmitters can be added within the SFN area without co-channel interferences even if the cells of the same frequency overlap. In fact, the multi-propagated SFN signals increase the performance of the network by producing SFN gain. In MFN mode, the frequency hand-over is performed each time the terminal moves from the coverage area of one site to another and no SFN gain is achieved in this mode. The most logical way to build up the DVB-H network is to use separate SFN isles covering e.g. single cities, so the practical network consists of both SFN and MFN solutions.

Especially in the SFN, the coverage planning is straightforward as long as the maximum distance of the sites does not exceed the allowed value defined by GI. This GI takes care of the safe reception of the multi-path propagated signals originated from various sites or due to the reflected radio waves. If the GI- and FFT-dependent geographical SFN boundary is exceeded, part of the sites starts acting as interferers instead of providing useful carrier.

The maximum size of the non-interfered SFN of DVB-H depends on the GI and FFT mode. The distance limitation between the extreme transmitter sites is thus possible to calculate in ideal conditions. Nevertheless, there might be need to extend the theoretical SFN areas e.g. due to the lack of frequencies.

Sites that are located within the SFN area minimise the effect of the inter-symbol interferences as GI protects the OFDM signals of DVB-H, although in some cases, sufficiently strong multi-path signals reflected from distant objects might cause interferences in tightly dimensioned network. On the other side, if certain degradation in the quality level of the received signal is accepted, it could be justified to even extend the SFN limits.

This section presents methods and results when estimating the level of SFN interference and the SFN gain. Reference [Pen08b] is used as a basis to investigate the effects of typical DVB-H radio parameters.

11.2.1   Controlled SFN Extension

According to [Dvb06], the GI and the FFT mode determinate the maximum delay that the mobile can handle for receiving correctly the multi-path components of the signals inside the SFN. Table 11.4 summarises the maximum allowed delays and respective distances. The maximum allowed distance per parameter setting has been calculated assuming that the radio signal propagates with the speed of light.

As long as the distance between the extreme transmitter sites is less than the safety margin dictates, the difference of the delays between the signals originated from different sites never exceeds the allowed value unless there is a strong multi-path propagated signal present.

The situation changes if the inter-site distance exceeds the allowed theoretical value. As an example, GI of $1/4$ and 8K mode provide 224 μs margin for the safe propagation delay. Assuming that the signal propagates with the speed of light, the SFN size limit is 300,000 km/s 224 μs yielding about 67 km of maximum distance between the sites. If any geographical combination of the site locations using the same frequency exceeds this maximum allowed distance, they start producing interference in those spots where the difference of the arriving signals is higher than 224 μs.

If the level of interference is greater than the noise floor, and the minimum C/N value that the respective mode requires in non-interfered situation is not any more obtained, the signal in that specific spot is interfered and the reception suffers from the frame errors that disturb the fluent following of the contents. In order to achieve

Table 11.4   The guard interval lengths and respective safety distances

GI	FFT $= 2$k	FFT $= 4$K	FFT $= 8$K
1/4	56 μs/16.8 km	112 μs/33.6 km	224 μs/67 km
1/8	28 μs/8.4 km	56 μs/16.8 km	112 μs/33.6 km
1/16	14 μs/4.2 km	28 μs/8.4 km	56 μs/16.8 km
1/32	7 μs/2.1 km	14 μs/4.2 km	28 μs/8.4 km

correct reception, the additional interference increases the required received power level of the carrier to $C/N \rightarrow C/(N + I)$. If the D_{eff}, i.e. the difference between the signals arriving from the sites, is more than the allowed safety distance in over-sized SFN, the site acts as an interferer. While the total carrier per interference and noise level (i.e. sum of carries per sum of interferences) complies with the minimum requirement for the C/N, the transmission is still useful.

Even if the $C/(N + I)$ level gets lower when the terminal moves from one site to another, the situation is not necessarily critical as the effective distance D_{eff} of the signals might be within the SFN limits e.g. in the middle of two sites, although their distance from each others would be greater than the maximum allowed. In other words, in certain locations, the otherwise interfering site might not be considered as interference in the respective spot, but it might give SFN gain instead by producing additional carrier C_2. This phenomenon can be observed in practice as the SFN interferences tend to accumulate primarily in the outer boundaries of the network.

Figure 11.8 shows the principle of the relative interference that increases especially when the terminal moves away from the centre of the SFN network.

When moving outside the network, the relative difference between the carrier and interfering signal gets smaller and it is thus inevitable that the $C/(N + I)$ will not be sufficient any more at some point for the correct reception of the carrier, although the C/N level without the presence of interfering signal would still be sufficiently high. The essential question is thus, where the critical points are found with lower $C/(N + I)$ value than the original requirement for C/N is, and where the interference thus converts active, i.e. when D_{eff} is longer than the safety margin.

As an example, the distance of two sites could be 70 km, which is more than D_{SFN} with any of the radio parameter combination of DVB-H. For the parameter set of FFT 8K and GI 1/8, the safety distance for D_{SFN} is about 34 km, which is clearly

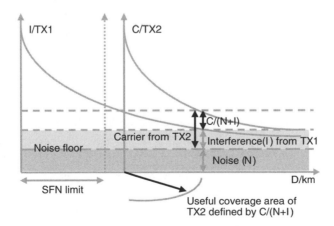

Figure 11.8 The principle of interference when the location of transmitter TX1 is out of the SFN limit

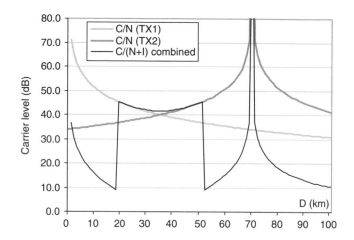

Figure 11.9 The combined $C/(N + I)$ along the route from 0 to 100 km, taking into account the SFN gain when the interference is not present

less than the distance of these sites. Let us define the radiating power (EIRP) for each site to be $+60$ dBm. We can now observe the received power level of the sites in the theoretical open area by applying the free space loss, f representing the frequency (in MHz) and d the distance (in km):

$$L = 20 \log f + 20 \log d + 32.44 \qquad (11.1)$$

Figure 11.9 shows the carrier (or interference) from TX1 located in 0 km and carrier (or interference) from TX2 located in 70 km, when the parameter set allows D_{SFN} of 34 km. The interference is included in those spots where the D_{eff} is higher than D_{SFN}. If the D_{eff} is shorter than D_{SFN}, the respective received useful power level is shown taking into account the SFN gain of these two sites by summing the absolute values of the power levels:

$$C_{\text{tot}} = \sqrt{C_1^2 + C_2^2} \qquad (11.2)$$

As Figure 11.9 shows, the TX1 is acting as a carrier and TX2 as interferer from 0 km (TX1 location) to 18 km, because the $D_{\text{eff}} > D_{\text{SFN}}$. Nevertheless, the carrier of TX1 is dominating within this area in order to provide sufficiently high $C/(N + I)$ for the successful reception for QPSK, CR $^1/_2$ and MPE-FEC $^1/_2$ as it requires 8.5 dB. The segment of 18–52 km is clear from the interferences as all the $D_{\text{eff}} < D_{\text{SFN}}$, and in addition, the receiver gets SFN gain from the combined carriers of TX1 and TX2. The TX1 starts to act as an interferer from 52 to 100 km (or, until the C/N limit of the used mode). Nevertheless, the interference of TX1 is already so attenuated such a far away from its origin that the $C/(N + I)$ is high enough for the successful reception

Table 11.5   The minimum C/N (dB) for the selected parameter settings

Parameters	C/N
QPSK, CR 1/2, MPE-FEC $^1/_2$	8.5
QPSK, CR 1/2, MPE-FEC 2/3	11.5
16-QAM, CR 1/2, MPE-FEC $^1/_2$	14.5
16-QAM, CR 1/2, MPE-FEC 2/3	17.5

from TX2 of above mentioned QPSK still within the area of 80–100 km. With any other parameter settings, the SFN interference level is high enough to affect the successful reception in these breaking points where the D_{eff} makes the signal act as interferer instead of carrier.

As can be seen from this example, the interference level takes place when the terminal moves towards the boundary sites of the network. As a result, the boundary site's coverage area gets smaller, and depending on the parameter setting, there will be interferences between the sites.

The required C/N for some of the most commonly used parameter setting can be seen in Table 11.5 [Dvb06]. The terminal antenna gain (loss) is taken into account in the presented values. Table 11.5 presents the expected C/N values in Mobile TU-channel (typical urban) for the 'possible' reference receiver.

The FFT size has impact on the maximum velocity of the terminal, and GI affects both the maximum velocity as well as the capacity of the radio interface. In fact, in these simulations, if only the requirement for the level of carrier is considered without the need to take into account the maximum functional velocity of the terminal or the radio channel capacity, the following parameter combinations result in the same C/N and $C/(N + I)$ performance due to their same requirement for the safety distances:

- FFT 8K, GI 1/4: only one set
- FFT 8K, GI 1/8: same as FFT 4K, GI 1/4
- FFT 8K, GI 1/16: same as FFT 4K, GI 1/8 and FFT 2K, GI 1/4
- FFT 8K, GI 1/32: same as FFT 4K, GI 1/16 and FFT 2K, GI 1/8
- FFT 4K, GI 1/32: same as FFT 2K, GI 1/16
- FFT 2K, GI 1/32: only one set.

11.2.2   SFN Simulations in Unlimited SFN Network

11.2.2.1   General

In order to estimate the error level of various sites caused by extending the theoretical geometrical limits of SFN network, a simulation can be carried out as presented in [Pen08c], [Pen08d] and [Pen08e]. For the simulation, the investigated variables can be e.g. the antenna height and power level of the transmitter, in addition to the GI and FFT mode that defines the SFN limits. This simulation method is relatively straightforward and suitable for the physical layer analysis. It takes into account the most important phenomena of the radio interface, including the long- and short-term fading via the mathematical basic formulas. The respective results indicate the rough levels what occur in the practical network without taking into account the in-depth link level events like the final performance of error coding methods. This can be assumed to reflect the reality sufficiently well when the respective requirement for the C/I levels is utilized. In more in-depth studies, a link layer simulator with the possibility to investigate the occurred bit errors and respective MPE-FEC correction capabilities should be used, but the presented simulator type gives good first-hand information about the overall impact of the radio parameter settings on the radio quality.

A block diagram of the presented SFN interference simulator is shown in Figure 11.10. In the case of [Pen08c], it produces the results to text files, containing the C/N, I/N and $C/(N + I)$ values showing the distribution in scale of -50 to $+50$ dB and with 0.1 dB resolution. The simulator places the terminal in 100×100 km^2 area according to the uniform distribution in function of the coordinates (x, y) during each simulation round. The other option is to create the network layout based on the treuse patterns with the respective size of K. The total C/I value is calculated per simulation round by observing the individual signals of the sites.

The nearest site is selected as a reference during the respective simulation round. If the arrival time delays difference $\Delta t_2 - \Delta t_1$ is less than that D_{SFN} defines, the respective signal is marked as a useful carrier C; otherwise it is marked as interference I.

The simulator calculates the expected radius of single cell in non-interfering case and fills the area with uniform cells according to the hexagonal model. This provides partial overlapping of the cells as showed in Figure 11.11.

Each simulation round provides information if that specific connection is useless e.g. if the criteria set of (1) effective distance $D_{eff} > D_{SFN}$ in any of the cells, and (2) $C/(N + I)$ < minimum C/N threshold. If both criteria are valid, and if the C/N would have been sufficiently high without the interference in that specific round, the SFN interference level is calculated.

Figure 11.11 shows an example of the site locations. As can be seen, the simulator calculates the optimal cell radius according to the parameter setting and locates the transmitters on map according to the hexagonal model, leaving ideal overlapping areas in the cell border areas. The size and thus the number of the cells depend on the radio parameter settings without interferences, and in each case, a result is a uniform service level in the whole investigated area. The same network setup is used throughout the complete simulation, and changed if the radio parameters of the following simulation require so.

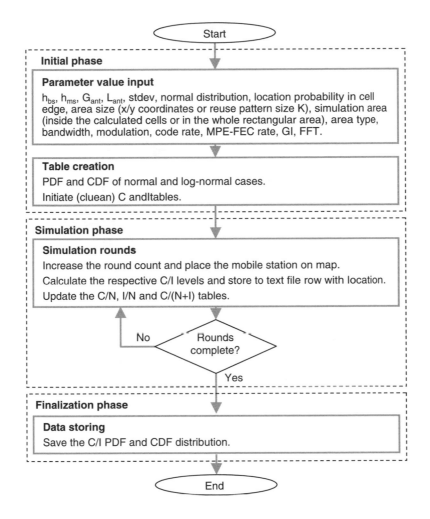

Figure 11.10 The simulator's block diagram

The behaviour of C and I can be investigated by observing PDF of the results. Nevertheless, the specific values of the interference levels can be obtained by producing a CDF from the simulation results.

Figure 11.12 shows two examples of the simulation results in CDF format when small and medium size city type was selected. In this specific case, the outage probability of 10% (i.e. area location probability of 90%) yields the minimum $C/(N + I)$ of 10 dB for 8K, which complies with the original C/N requirement (8.5 dB) of this case. On the other hand, the 4K mode results in approximately 7 dB with 10% outage, which means that the minimum quality targets cannot quite be achieved any more with these settings.

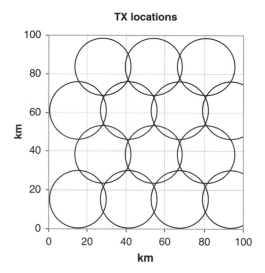

Figure 11.11 An example of the transmitter site locations the simulator has generated

When the total carrier per interference levels are estimated, the sum of the useful carriers and sum of interferences can be calculated separately by the following formulas, using the respective absolute power levels (W) for the C and I components:

$$C_{\text{tot}} = \sqrt{C_1^2 + C_2^2 + \cdots + C_n^2}$$

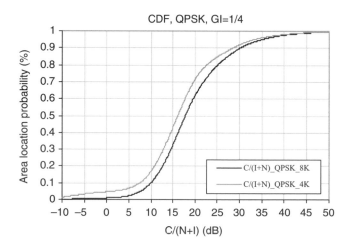

Figure 11.12 An example of the cumulative distribution of $C/(N + I)$ for QPSK 4K and 8K modes with antenna height of 200 m and Ptx $+60$ dBm

$$I_{\text{tot}} = \sqrt{I_1^2 + I_2^2 + \cdots + I_n^2}$$

As presented in [Pen08c], in each simulation round, the mobile is placed to new location on the field and the site with the highest field strength is identified and selected as a reference. The simulator then investigates the propagation delays of signals between the nearest and each one of the other sites, and calculates if the difference between these arriving signals. The signal is marked as interferer if the effective difference of the arrival time of these two signals is greater than the SFN limit D_{SFN}, otherwise the signals are assumed to give SFN gain and they are summed as useful carriers. The same procedure is repeated for each arriving signal that just appears stronger than the noise level is.

The total path loss for each arriving signal can be calculated by applying the following formula:

$$L_{\text{tot}} = L_{\text{pathloss}} + L_{\text{norm}} + L_{\text{other}} \qquad (11.3)$$

L_{pathloss} can be estimated by applying Okumura–Hata, ITU-R P.1546 or any other suitable prediction model. L_{norm} represents the fading loss caused by the long-term variations, and other losses may include e.g. the fast fading as well as antenna losses.

For the long-term fading, a normal distribution can be applied as is typically done in the related analysis. This creates the practical variations for the signal level. The PDF of the long-term fading is the following:

$$\text{PDF}(L_{\text{norm}}) = \frac{1}{\sqrt{2\pi}\sigma} \exp\left[\frac{-(x-x)^2}{2\sigma^2}\right]$$

The term x represents the loss value, andis the average loss (0 in this case). In the snapshot-based simulations, the L_{norm} is calculated for each arriving signal individually as the different events do not have correlation. The respective PDF and CDF are obtained by creating a probability table for normal distributions. Figure 11.13 shows an example of the PDF and CDF of normal distributed loss variations, when the mean value is 0 and standard deviation is 5.5 dB.

The fast (Rayleigh) fading is present in those environments where multi-path radio signals occurs e.g. on the street level of cities. It can be presented with the following PDF:

$$L_{\text{log norm}} = \frac{x}{\delta^2} e^{-\left(\frac{x^2}{2\delta^2}\right)}$$

Figure 11.14 shows the PDF and CDF of the fast fading representing the variations of short-term loss when the standard deviation is set to 5.5 dB.

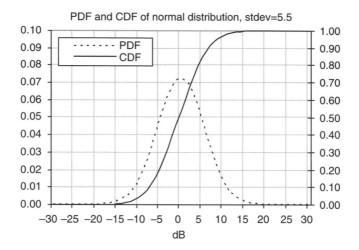

Figure 11.13 PDF and CDF of the normal distribution representing the variations of long-term loss when the standard deviation is set to 5.5 dB

11.2.2.2 Results

By applying the principles of the DVB-H simulator, the *C/I* distribution can be obtained according to the selected radio parameters. In this example, the variables were the modulation scheme (QPSK and 16-QAM), antenna height (20–200 m) and FFT mode (4K and 8K).

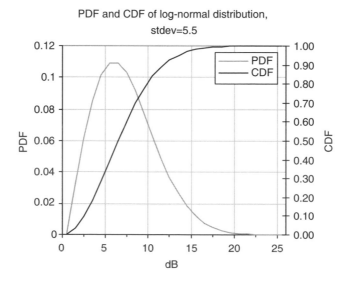

Figure 11.14 PDF and CDF of the log-normal distribution for fast fading

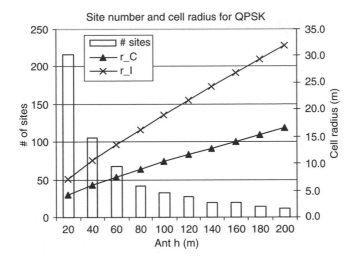

Figure 11.15   The network dimensions for the QPSK simulations

Figures 11.15 and 11.16 show the resulting networks that were used as a basis for the simulations. The simulator selects randomly the mobile terminal location on the map and calculates the *C/I* that the network produces at that specific location and moment. This procedure is repeated for required number of times. In this case, 60,000 simulation rounds were used. One of the results after the complete simulation is the estimation for the occurred errors due to the interfering signals from the

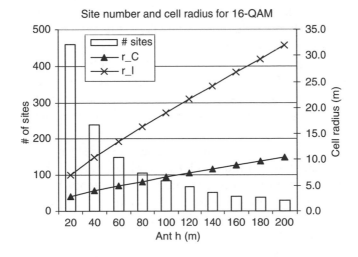

Figure 11.16   The network dimensions for the 16-QAM simulations

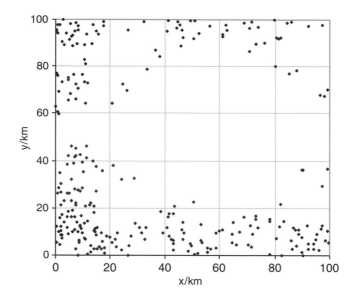

Figure 11.17 An example of the results in geographical format with $C/(N + I) < 8.5$ dB

sites exceeding the functional SFN distance (i.e. if the arrival times of the signals exceed the maximum allowed delay difference).

In Figure 11.17, the plots indicate the locations where the results of $C/(N + I)$ corresponds to 8.5 dB or less for QPSK. In this case, the interfering plots represent the relative SFN area error rate (SAER) of 0.83%, i.e. the erroneous (SAR) cases over the number of total simulation rounds as for the simulated plots.

It can be assumed that when the SER level is sufficiently low, the end-users will not experience remarkable reduction in the DVB-H reception due to the extended SFN limits. In this analysis, an SER level of 5% is assumed to still provide with sufficient performance as it is in line with the limits defined in [Dvb06] for frame error rate before the MPE-FEC (FER) and frame error rate after the MPE-FEC (MFER). The nature of the SER is slightly different, though, as the interferences tend to cumulate to certain locations as can be observed from Figure 11.17 obtained from the simulator.

According to the simulations, the SFN interference level varies clearly when the radio parameters are tuned. Figures 11.18–11.21 summarise the respective SFN area error analysis, the variable being the transmitter antenna height. These figures show that with the uniform radio parameters and varying the antenna height, modulation and FFT mode, the functional settings can be found regardless the exceeding of the theoretical SFN limits.

If a 5% limit for SER is accepted, the analysis show that antenna height of about 80 m or lower produces SER of 5% or less for QPSK, 8K, with minimum

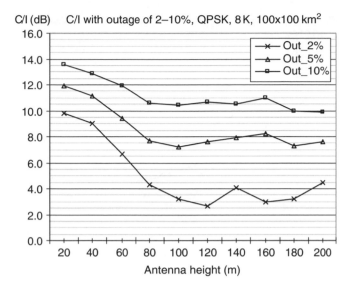

Figure 11.18 The summary of the case 1 (QPSK, 8K). The results show the $C/(N + I)$ with 2%, 5% and 10% SER criteria

$C/(N + I)$ requirement of 8.5 dB. If the mode is changed to 4K, the antenna height should be lowered to 35–40 m from ground level in order to comply with 5% SER criteria in this very case. 16-PSK produces higher capacity and smaller coverage.

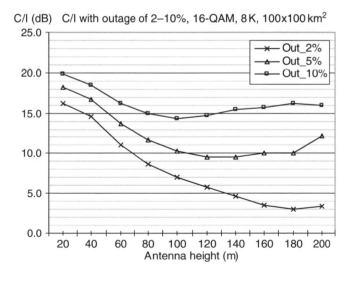

Figure 11.19 The summary of the case 2 (16-QAM, 8K)

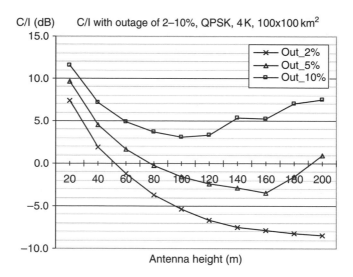

Figure 11.20   The summary of the case 3 (QPSK, 4K)

The results are showing this clearly as the respective SER of 5% (16-QAM, 8K and minimum $C/(N+I)$ requirement of 14.5 dB for this modulation) allows the use of antenna height of about 120 m. If the mode of this case is switched to 4K, the antenna should be lowered down to 50 m in order to still fulfil the SER 5% criteria.

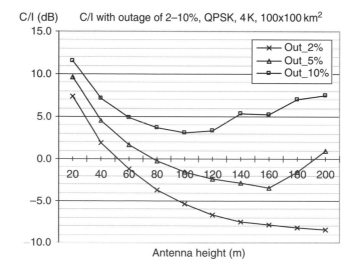

Figure 11.21   The summary of the case 4 (16-QAM, 4K)

The +60 dBm EIRP represents relatively low power. The higher power level raises the SER level accordingly. Thus for the mid- and high-power sites, the optimal setting depends even more on the combination of the power level and antenna height. According to these results, it is clear that the FFT mode 8K is the only reasonable option when the SER should be kept in acceptable level. Especially the QPSK modulation might not allow easy extension of SFN as the modulation provides largest coverage areas. On the other hand, when providing more capacity, 16-QAM is the most logical solution as it gives normally sufficient capacity with reasonable coverage areas. The stronger CR and MPE-FEC rate decreases the coverage area but it is worth noting that the interference propagates equally also in those cases.

The general problem of the SER arises from the different loss behaviour of the useful carrier and interfering signal. Depending on the case, the interfering signal might propagate two to three times further away from the originating site compared to the useful carrier.

In practice, the SER level can be further decreased by minimising the propagation of the interfering components. This can be done e.g. by adjusting the transmitter antenna down-tilting and using narrow vertical beam widths, thus producing the coverage area of the carrier and interference as close to each others as possible. Also the natural obstacles of the environment can be used efficiently for limiting the interferences far away outside the cell range.

Figure 11.22 shows the previously presented results presenting the outage percentage for the different modes having 8.5 dB $C/(N + I)$ limit for QPSK and 14.5% for 16-QAM cases in relation to transmitter antenna height.

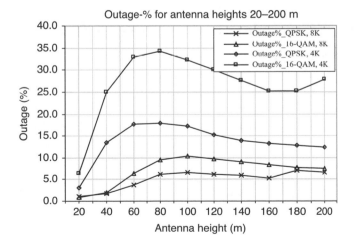

Figure 11.22 The summary of the cases 1–4 presenting the outage percentage related to the transmitter antenna height

11.2.2.3   Conclusion

The presented simulation method provides both geographical and cumulative distribution of the SFN gain and interference levels. The method takes into account the balancing of the coverage and capacity as well as the optimal level of SFN gain and the interference level in case the over-sized SFN is used. It can be applied for the theoretical e.g. hexagonal cell layouts, as well as for the practical environments, taking into account the radio propagation modelling for different sites. The method can thus be used in the detailed optimization of the DVB-H networks. The principle of the simulator is relatively straightforward and the method can be applied by using various different programming languages.

The results of [Pen08c], [Pen08d] and [Pen08e] show that the radio parameter selection is essential in the detailed planning of the DVB-H network. As the graphical presentation of the results indicates, the effect of the parameter value selection on the interference level, and thus on the quality of service, can be drastic, which should be taken into account in the detailed planning of DVB-H SFN.

Especially the controlled extension of the SFN limit might be an interesting option for the DVB-H operators. The simulation method and related results show the logical behaviour of the SFN error rate when varying the essential radio parameters. The results also show that the optimal setting can be obtained using the respective simulation method by balancing the SFN gain and SFN errors. As expected, the 8K mode is most robust when extending the SFN while 4K limits the maximum site antenna height. 16-QAM provides suitable performance for the extension, but according to the results, even QPSK that provides larger coverage areas is not useless in SFN extension when selecting the parameters correctly.

11.3   Effect of Transmitter Power on Network Costs

The DVB-H capacity and coverage can be achieved by many different combinations of the parameter values, including the variation of the transmitter power levels and antenna heights. Taking into account the limitations of SFN interferences, the maximum coverage can be achieved technically by locating the transmitter antennas as high as possible and by using maximum transmitter power levels. Nevertheless, in detailed optimization, also the cost of different solutions should be taken into account.

As an example, more output power the transmitter provides, higher will be the equipment complexity and power consumption, affecting OPEX of the network. In optimal deployment of the network, it is thus essential to identify the most relevant technical parameters and investigate their impacts on the initial, i.e. CAPEX, and OPEX. Even in case of relatively small DVB-H network deployment, the proper election of the transmitter types (power levels) might considerably reduce the final costs of the network.

The DVB-H planning is done for core and radio sub-systems. For both parts, the proper capacity is dimensioned taking into account the short-term operation and preferably mid-term evolution prediction.

The capacity of the DVB-H network is calculated by taking into account the modulation scheme (QPSK, 16-QAM or 64-QAM) and error correction scheme (CR of 1/2, 2/3, 3/4, 5/6 or 7/8). As a difference with DVB-T, there is an enhanced correction method, MPE-FEC, defined in DVB-H, which provides additional protection against the effects of multi-path propagation and impulse noise in mobile environment. This parameter can be set with the values of 1/2, 2/3, 3/4 and 5/6. Furthermore, the final design of the network depends on the value for the GI, interleaver mode (2K, 4K, 8K), area location probability (typically between 70% and 95%), shadowing margin and possibly SFN gain. Depending on these settings, the balance between capacity and coverage can be found.

The most important initial network planning input is normally the required capacity in radio interface, which dictates the adequate transmitter power level. The main limitations depend on the general regulations for the RF radiation (including the EMC and human exposure limits) as well as on the practical issues since the cost of different types of transmitters is not linear in function of the power levels. Another significant factor is the antenna height, which does have an impact on the DVB-H coverage.

As the CAPEX and OPEX are considered, there are various other aspects that affect the final cost of the network. Also the amount of leased or own items, like transmission lines, transmitter sites, etc., does have its impact. The cost depends mainly on the transmitter equipment complexity, transmission lines, transmitter sites, towers or rooftops, antenna feeders and antennas. The detailed cost list might include the material that is needed for the installation of the equipment. In addition, the in-depth cost optimization should take into account the installation services and other immaterial items like the cost of the planning, preparation of the site drawings, site acquisition, license fees, maintenance costs, etc.

11.3.1   Description of the Methodology

In order to identify the initial optimal parameter values when minimising the CAPEX and OPEX of the DVB-H networks, a systematic methodology can be applied. The process contains the following high-level revision as a basis for the investment decision:

- Identify the main items that affect on CAPEX and OPEX.
- Estimate the cost for each item.
- Calculate the total CAPEX and (yearly) OPEX for single transmitter site.
- Calculate the single cell radius for each case, averaging the investigated area as a uniform type, or selecting a set of separate uniform area types (that are calculated individually), using adequate RF propagation model.

- Select sufficiently large service area and estimate by using e.g. hexagonal model, how many sites there should be obtained in each case in order to cover the area with the uniform quality of service level.
- Calculate the total CAPEX and OPEX for each case for comparison purposes.
- For the accurate estimation, it is important to select all the major items that has cost impact on the network, and as many minor details as is seen found reasonable. In this approach, the transmitter power level is selected as the variable while the core network with source signals, capacity, bit rate per channel, etc. can be assumed to be the same in each case.

11.3.2 CAPEX and OPEX Estimation

For the realistic DVB-H CAPEX estimation, the following cost items can be taken into account:

- transmitters;
- antenna systems (with antenna feeders, power splitters, jumpers and connectors);
- other materials, like feeder brackets, tools, etc.;
- transmitter site acquisition and preparation work;
- transmitter, antenna system and other material installation and commissioning work

For the OPEX, the longer term items include at least the following:

- transmission (leased lines, satellite transmission, etc.);
- maintenance of the transmitters, other equipment and site;
- tower and/or site rent;
- electricity consumption

The transmission has a key role in the OPEX per site. Depending on the needed capacity, the technical solution can be done by using e.g. sufficient amount of E1/T1 lines, or implementing fibre optics that provide sufficiently wide bit pipe. For the remote areas with relatively large proportion of sites difficult to access, a satellite transmission might provide with optimal solution. The basic cost of the satellite transmission is normally clearly higher than in terrestrial cable solution, but the single satellite link usually covers all the needed sites.

For the electricity consumption, a rough estimation of six times the output power level can be used in this analysis unless the practical values are available. For the other items, the costs depend on the equipment and service provider list prices, taking into account the possible volume and other discounts. Furthermore, the prices can be estimated separately for the main transmitter sites and gap-fillers.

In order to simplify the calculation, it is sufficient to take into account the number of main transmitter sites, and possibly estimating a lump cost for the gap-filler sites. The number and transmitting powers of gap-fillers depend on the wanted level of the outdoor and indoor coverage areas.

The CAPEX and OPEX include both common costs as well as the costs that depend on the type of the transmitter site. The common CAPEX estimation includes the site acquisition and structural analysis, whereas the transmitter, antenna system, power splitter, brackets and installation work depend on each site type.

The common OPEX items include the transmission lease (assuming the same bit pipe is delivered to each site), and average tower or rooftop site rent. The other OPEX items depend on the site type, and include e.g. electricity consumption and maintenance work, which both depend on the transmitter type.

For the CAPEX, the transmitter cost plays a key role. As the power level of the equipment gets higher, the relative price of the power (W) gets normally lower. This is logical as the equipment contains common parts, design work, racks, etc. that generates equal type of costs regardless of the differences in the power amplifier block. On the other hand, the highest power level transmitters are more complicated with e.g. liquid cooling requirements that raise the cost as the power level is considered. Figure 11.23 summarises an example of one snap-shot analysis of market prices of the DVB-H transmitters. The analysis shows the relative cost behaviour against the transmitter power level of that specific transmitter vendor.

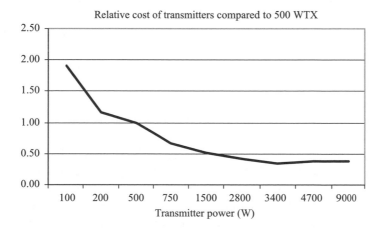

Figure 11.23   An example of the relative comparison of the DVB-H transmitter power level price, i.e. the cost of the single watt produced by different transmitter types. The values represent a small snap-shot of the market, which obviously changes during the time and depends on the market movements. The values should thus be interpreted as an example

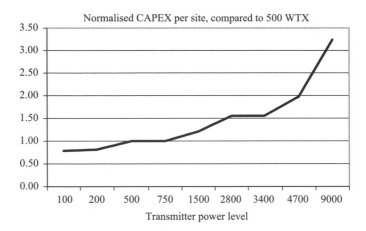

Normalised CAPEX per site, compared to 500 WTX

Figure 11.24 An example of the CAPEX per site. The values are normalised using the 500-W transmitter as a reference. The costs include the transmitter and antenna system as well as related installation services

In Figure 11.23, the 500-W transmitter represents the normalised reference (i.e. the price of single watt produced by 500-W transmitter). According to this specific example, the cost for producing a single watt is half when using the 1500-W transmitter instead of 500 W type (i.e. in this specific example, three 500 W solution is two times more expensive compared to one 1500 W). Note that the cost values depend totally on the vendor list prices and that Figure 11.23 gives only an idea of how the price of single watt produced by different transmitter types could possibly behave.

A case study can now be performed based on the above-mentioned cost items. The CAPEX estimation per site for this case study can be seen in Figure 11.24. Again, the values are shown by normalizing and using the 500-W case as a reference. The figure gives a rough idea of how much the total transmitter site would cost when the transmitter power level (the transmitter type) is varied. As an example, when using 4.7-kW transmitter type instead of 500-W transmitter, the site cost would double in this example.

When taking into account the operating costs of the sites as well, Figure 11.25 can be produced. It shows the OPEX for the previously presented transmitter types.

As can be observed from Figure 11.25, the major OPEX items are basically constant except for the electricity consumption. For the highest power level cases, this might turn out to be an important cost and thus has negative impact if the network is planned with only few high-power sites. For the transmission, the terrestrial cable solution with sufficiently high bit pipe was used commonly in each case of this analysis. The other solutions like satellite transmission result in logically different cost structures that should also be taken into account.

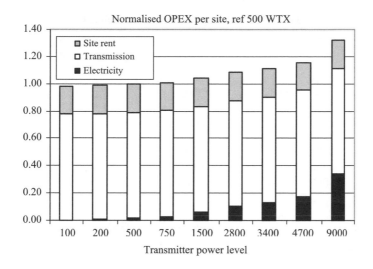

Figure 11.25   The relative behaviour of OPEX, with respective analysis of the OPEX items

11.3.3   Coverage Estimation

In the initial planning phase, an Okumura–Hata propagation model can be used to estimate the radius of a single cell in each transmitter case,. In this case analysis, a sub-urban area type was used with QPSK modulation, antenna height of 100 m, terminal height of 1.5 m, service level of 90% (area location probability), shadowing margin of 5.5 dB, frequency of 700 MHz and receiver antenna gain of 7 dBi. The CR of $\frac{1}{2}$ and MPE-FEC of $\frac{3}{4}$ was selected. The parameter selection yields the minimum received power level of about −87 dBm for the functional service.

The cell radius was calculated for the transmitter output power levels of 100–9000 W. Antenna gain of 13 dBi was used in each case, which is the result of directional antennas (or antenna arrays, depending on the power level) installed in three sectors without down-tilting.

The calculation should take into account the jumper, connector, power splitter and feeder losses. A feeder of 1 5/8″ with the loss of 1.9 dB/100 m was selected for the power levels of 100–3400 W, and a 3″ cable with the loss of 1.5 dB/100 m was used for the power levels of 4700–9000 W. An estimation of 10% for transmitter filter loss was used in each case. The assumption was to use the antenna in tower, which means that the same antenna feeder length (133 m) was used in each case. The cell range that was obtained by taking into account the above-mentioned values can be seen in Table 11.6.

The next step of this methodology includes the selection of a physical service area. The cells are then placed in the planned area in order to estimate the total cost of the network for each power level case. The area is thus filled with cells as tightly as possible, using the hexagonal model.

Table 11.6 The calculated cell range for each case

TX power	EIRP (w)	Range r (km)
100	576	6.4
200	1152	8.0
500	2880	10.7
750	4321	12.1
1500	8641	15.1
2800	16130	18.4
3400	19587	19.5
4700	31019	20.7
9000	59398	22.6

The cell coverage area is represented with a circle that touches the edges of the hexagonal element. The circles overlap partially in the cell edges resulting in relatively realistic presentation of the coverage areas. In practice, this provides the service continuity as well as SFN gain due to the multiple path of the radio signal. The overlapping area presented in this analysis can be calculated geometrically by comparing the surface of the circle with the hexagonal area.

When the radius of the cell (circle) is r, the surface of the hexagonal inside of the circle is

$$Ah = 6r \cos(30)\frac{r}{2} = 3r^2\cos(30)$$

In this analysis, the overlapping area is taken into account as a reduction factor Rf when estimating the total cell number in given service area. It means that the overlapping area exists in the investigated area, but the reduction factor gives the possibility to calculate the single cell areas and the number of the cells by using the formula of the surface of the circles. The reduction factor can be obtained by the following formula:

$$Rf = \frac{Ah}{Ac} = \frac{3r^2\cos(30)}{\pi r^2} = \frac{3\cos(30)}{\pi} \approx 82.7\%$$

With the reduction factor, it is thus possible to estimate how many partially overlapping omni-cells (N_{cells}) with a form of the circle and the radius of r fits into the planned service area. The formula is the following:

$$N_{cells} = \frac{A_{tot}}{A_{cell}} Rf = \frac{A_{tot}}{\pi r_{cell}^2} Rf$$

11.3.4 Results

When observing the results calculated for the total service area (in this analysis an area of $100 \times 100\,\text{km}^2$ was selected), the following CAPEX and OPEX relation can be obtained depending on the power level of the site.

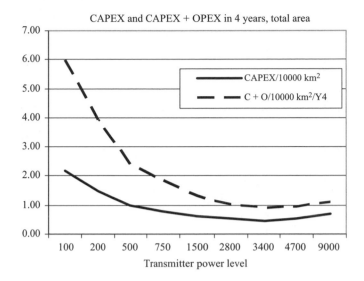

CAPEX and CAPEX + OPEX in 4 years, total area

Transmitter power level

Figure 11.26 The relative CAPEX and OPEX comparison of different power level cases

As can be seen from Figure 11.26, there exists an optimal point for both CAPEX and OPEX curves. In this specific case, the optimal power level is found in 3.4-kW category.

It is interesting to notice that the OPEX and CAPEX curves follow the general trend of the power unit price for different transmitter types, but nevertheless, the final relative CAPEX and combined CAPEX/OPEX grow faster for the highest power level cases that take into account all the relevant cost items for each power level case.

For the OPEX, a more specific analysis can still be done. Figure 11.27 shows the development of the cumulative operating costs during 4 years from the initial deployment of the network. The yearly OPEX is constant for each transmitter case, producing lines with certain angular coefficient.

It can be seen that e.g. for the power level of 750 W, the initial cost is lower than in average with the small-power levels, but the operating cost of 750-W case is considerably higher than could perhaps be expected. In this particular case, it can also be seen that the highest power level, i.e. 9 kW, is a relatively expensive solution as the CAPEX is considered, and regardless of the considerably lower amount of the sites compared to the lower power level cases, the cumulative OPEX development (angular coefficient of the line) is only slightly lower than that of the mid-power transmitters, mainly because of the higher power consumption.

When observing the angular coefficients of each case and taking into account the development of the network for 4 years, the optimal power level is thus found in the mid-level power range, i.e. the respective transmitters provide with the lowest CAPEX and OPEX of the DVB-H network in this specific case.

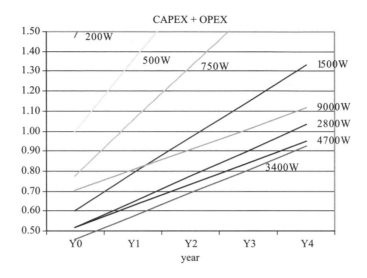

Figure 11.27   An analysis of the OPEX development

In generic situation, the coefficients can be calculated by the following formula:

$$k = \frac{y - y_0}{(x - x_0)}$$

Here y_0 represents the CAPEX (the cost in initial year), and x_0 can be marked as 0 as it represents the beginning of the operation, i.e. the year 0. It is thus straightforward to calculate the total cost of the network after x years:

$$y = kx + y_0$$

The coefficients of this specific analysis can be seen in Table 11.7. It can be noted that the coefficient always lowers when the transmitter power level is higher. The

Table 11.7   The angular coefficient of different transmitter power levels

TX_{power}	Y_0	k
100	2.20	0.95
200	1.47	0.61
500	1.00	0.35
750	0.77	0.27
1500	0.60	0.18
2800	0.52	0.13
3400	0.46	0.12
4700	0.52	0.11
9000	0.70	0.10

task would thus be to find the case that yields lowest total cost (CAPEX and cumulated OPEX) within x years, i.e. within the expected operating period of the network.

In this specific analysis, the 3400-W transmitter would provide the lowest total cost for the time scale of 0–6 years of operation. The 4700 W turns out to be more attractive if the network would operate during 7–45 years, and theoretically, the 9000-W transmitter would yield the lowest costs if the network would operate in the very same setup for at least 46 years. This type of time periods are of course theoretical as the equipment should probably be updated or renewed anyway time to time, but this methodology gives a rough idea about the possible return on investment times per studied case.

12

Future

The cultures all over the world have adapted to the new modern multimedia era. The time and location independent mobile communications with voice calls and text messaging can already be considered as basic services, and more advanced multimedia applications via high-speed mobile data transmission methods are becoming a part of our normal life style.

One of the important aspects of the information society is the possibility of seeking and finding the information in real time regardless of the location. The DVB provides a good basis for the digital broadcast communications via its handheld version DVB-H. With its return channel e.g. via GPRS, it can be used for various interactive, real-time handling of the information. Some of the logical applications of DVB-H are informative audio/video clips, as well as entertainment programs. Not only the usual news is interesting to receive via DVB-H, but the system also provides an excellent base for delivering information about exceptional circumstances. DVB-H can be used as an extra communication channel to warn and advice people e.g. about sudden weather changes, nature disasters and other potential catastrophes that require immediate actions. Some common examples might be the real-time hurricane warnings, tsunami-related information, etc. In addition, as the DVB-H can deliver both local and wider contents, it is a logical piece in the set of communication methods in order to inform about traffic jams and local accidents, etc.

As DVB-H is strongly in liaison with the mobile communications system as the GSM and UMTS are often integrated in the same channel, it provides good methods for the interactions. As an example, televoting could be inserted as a temporal service via the DVB-H program with corresponding internet link or other means to

The DVB-H Handbook   Jyrki T.J. Penttinen, Erkki Aaltonen, Jani Väre and Petri Jolma
© 2009 John Wiley & Sons, Ltd

obtain more information by establishing a dedicated link between the user and service provider.

The first DVB-H specifications have been ready since 2004, for bringing broadcast services to battery-powered handheld receivers. The specification set was formally adopted as an ETSI standard in November 2004, and ever since pilots and commercial networks have been constructed. The first version of the DVB-H is proven to work as designed in moving environment. Its radio interface has been optimized for the difficult characteristics of the radio interface, giving possibility of using the service in varying speeds and in different environments both in outdoors and indoors.

The DVB-H has its evolution paths as has been seen to be the case in basically every modern telecommunications technologies. The next step of DVB-H is called DVB-NGH, which indicates the planning of the next generation of the mobile broadcast system. It is an enhanced version of the first DVB-H, resulting e.g. in a more optimized coding scheme.

The satellite component that can be used on the road has been missing in the DVB family. The DVB-SH is being planned in order to cover large areas in the coverage areas of the satellite signal. Published by ETSI in March 2008, the DVB-SH specification is designed to enable the delivery of mobile TV services in S-band over hybrid satellite and terrestrial networks.

The broadcast service type is attractive as one piece in the complete telecommunication market as the current mobile networks have not yet reached the sufficient capacity for the delivery of respective contents. Figure 12.1 clarifies the future expectations of the uncast vs. broadcast markets.

One of the possibilities is that the uncast type of traffic increases as the capacity of the evolved networks gets higher. This might have negative impact on the mobile

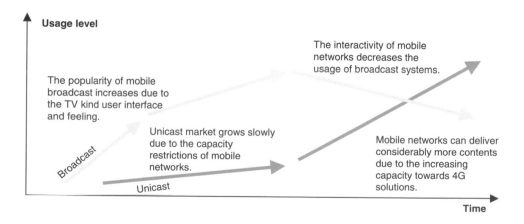

Figure 12.1   The future expectations of the broadcast markets

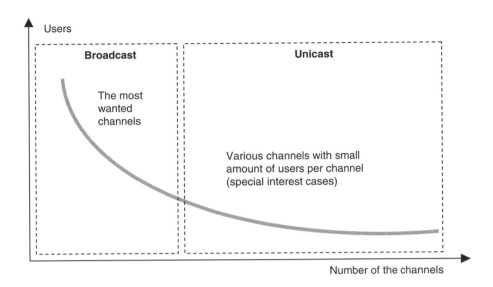

Figure 12.2   The principle of the division of broadcast and unicast channel types

broadcast usage because it has limited amount of channels. In fact, one of the visions might be that the broadcast remains offering the most popular contents for the big audience and with enhanced interactivity via the evolved networks. For the rest of the channels, a combined unicast service can be used for offering a big amount of different contents via more advanced interactivity. For the rest of the channels, a combined unicast service can be used for offering a big amount of different contents via more advanced interactivity as shown in Figure 12.2.

In general, DVB-H type of mobile broadcast is suitable for the delivery of around 15–30 A/V channels with characteristics of mobile TV content for the big audience, depending on the network parameter settings and quality level. Unicast is suitable for the user-controlled personalized channels, with higher level of interactivity as DVB-H with mobile return channel offers. In addition, MBMS can deliver multimedia for communities, although it reserves a part of the 3G radio network capacity and cannot thus be used in wider scale until 4G capacities.

In general, Mobile TV and IPTV will converge little by little with other systems, enabling new user experiences. The content of mobile TV and fixed IPTV will deliver the same contents for mobile screens as well as fixed household television sets. This provides with the means of optimizing the services, and gives the customers a one feel type of experiences. The usage of the services will be finally similar whatever is the terminal type, in moving and fixed environments, including the return channel services like televoting, electrical purchase, etc.

13

APPENDIX 1: DVB Standards List

13.1 Transmission

EN 300 421 V1.1.2 (08/97). Framing structure, channel coding and modulation for 11/12 GHz satellite services.

TR 101 198 V1.1.1 (09/97). Implementation of Binary Phase Shift Keying (BPSK) modulation in DVB satellite transmission systems.

EN 302 307 V1 .1.2 (06/06). Second generation framing structure, channel coding and modulation systems for broadcasting, interactive services, news gathering and other broadband satellite applications.

TR 102 376 V1.1.1 (02/05). User guidelines for the second generation system for broadcasting, interactive services, news gathering and other broadband satellite applications.

TS 102 441 V1.1.1 (10/05). DVB-S2 adaptive coding and modulation for broadband hybrid satellite dialup applications.

EN 300 429 V1.2.1 (04/98). Framing structure, channel coding and modulation for cable systems.

EN 300 473 V1.1.2 (08/97). DVB Satellite Master Antenna Television (SMATV) distribution systems.

TS 101 964 V1.1.1 (08/01). Control channel for SMATV/MATV distribution systems; baseline specification.

TR 102 252 V1.1.1 (10/03). Guidelines for implementation and use of the control channel for SMATV/MATV distribution systems.

EN 300 744 V1.5.1 (11/04). Framing structure, channel coding and modulation for digital terrestrial television.

The DVB-H Handbook Jyrki T.J. Penttinen, Erkki Aaltonen, Jani Väre and Petri Jolma
© 2009 John Wiley & Sons, Ltd

TR 101 190 V1.2.1 (11/04). Implementation guidelines for DVB terrestrial services; transmission aspects.

TS 101 191 V1.4.1 (06/04). Mega-frame for Single Frequency Network (SFN) synchronization.

DVB BlueBook A122 (06/08). Frame structure channel coding and modulation for a second generation digital terrestrial television broadcasting system (DVB-T2).

EN 302 304 V1.1.1 (11/04). Transmission system for handheld terminals.

TR 102 377 V1.2.1 (11/05). Implementation guidelines for DVB handheld services.

DVB BlueBook A092r2 (07/07). Implementation guidelines for DVB handheld services (draft TR 102 377 V1.3.1).

TR 102 401 V1.1.1 (05/05). DVB-H Validation Task Force report.

TS 102 585 V1.1.1 (07/07). System specifications for satellite services to handheld devices (SH) below 3 GHz.

EN 302 583 V1.1.1 (07/07). Framing structure, channel coding and modulation for satellite services to handheld devices (SH) below 3 GHz.

A120 (dTS 102 584 V1.1.1) (05/08). DVB-SH implementation guidelines.

DVB BlueBook A131 (11/08). MPE-IFEC.

EN 300 748 V1.1.2 (08/97). Multipoint Video Distribution Systems (MVDS) at 10 GHz and above.

EN 300 749 V1.1.2 (08/97). Framing structure, channel coding and modulation for MMDS systems below 10 GHz.

EN 301 701 V1.1.1 (08/00). OFDM modulation for microwave digital terrestrial television.

EN 301 210 V1.1.1 (02/99). Framing structure, channel coding and modulation for Digital Satellite News Gathering (DSNG) and other contribution applications by satellite.

TR 101 221 V1.1.1 (03/99). User guidelines for Digital Satellite News Gathering (DSNG) and other contribution applications by satellite.

EN 301 222 V1.1.1 (07/99). Co-ordination channels associated with Digital Satellite News Gathering (DSNG).

13.2 Multiplexing

EN 300 468 V1.8.1 (10/07). Specification for Service Information (SI) in DVB systems.

DVB BlueBook A038r4 (07/08). Specification for Service Information (SI) in DVB systems (draft EN 300 468 V1.9.1).

TR 101 211 V1.8.1 (08/07). Guidelines on implementation and usage of Service Information (SI).

Draft TR 101 162 V1.2.1 (02/01). Allocation of Service Information (SI) codes for DVB systems.

EN 300 472 V1.3.1 (05/03). Specification for conveying ITU-R System B Teletext in DVB bitstreams.

EN 301 775 V1.2.1 (05/03). Standard for conveying VBI data in DVB bitstreams.

TS 102 823 V1.1.1 (11/05). Carriage of synchronised auxiliary data in DVB transport streams.

EN 301 192 V1.4.2 (04/08). Specification for data broadcasting.

TR 101 202 V1.2.1 (01/03). Specification for data broadcasting; Guidelines for the use of EN 301 192.

TS 102 006 V1.3.2 (07/08). Specification for system software update in DVB systems.

TS 102 323 V1.3.1 (04/08). Carriage and signalling of TV-Anytime information in DVB transport streams.

TS 102 606 V1.1.1 (10/07). Generic Stream Encapsulation (GSE) protocol.

TS 102 833 V1.1.1 (11/08). File format specification for the storage and playback of DVB services.

13.3   Source Coding

TS 101 154 V1.8.1 (07/07). Implementation guidelines for the use of MPEG-2 systems, video and audio in satellite, cable and terrestrial broadcasting applications.

TS 102 154 V1.2.1 (05/04). Implementation guidelines for the use of MPEG-2 systems, video and audio in contribution applications.

TS 102 005 V1.3.1 (07/07). Implementation guidelines for the use of audio-visual content in DVB services delivered over IP.

13.4   Subtitling

EN 300 743 V1.3.1 (11/06). Subtitling systems.

13.5 Interactivity

ETS 300 802 V1 (11/97). Network-independent protocols for DVB interactive services.

TR 101 194 V1.1.1 (06/97). Guidelines for implementation and usage of the specification of network independent protocols for DVB interactive services.

ES 200 800 V1.3.1 (11/01). Interaction channel for Cable TV distribution systems (CATV).

TR 101 196 V1.1.1 (12/97). Interaction channel for Cable TV distribution systems (CATV); Guidelines for the use of ETS 300 800.

ETS 300 801 V1 (08/97). Interaction channel through Public Switched Telecommunications Network (PSTN)/Integrated Services Digital Networks (ISDN).

EN 301 193 V1.1.1 (07/98). Interaction channel through the Digital Enhanced Cordless Telecommunications (DECT).

EN 301 199 V1.2.1 (06/99). Interaction channel for Local Multipoint Distribution System (LMDS) distribution systems.

TR 101 205 V1.1.2 (07/01). Guidelines for the implementation and usage of the DVB interaction channel for Local Multipoint Distribution System (LMDS) distribution systems.

EN 301 195 V1.1.1 (02/99). Interaction channel through the Global System for Mobile Communications (GSM).

TR 101 201 V1.1.1 (10/97). Interaction channel for Satellite Master Antenna TV (SMATV) distribution systems; Guidelines for versions based on satellite and coaxial sections.

EN 301 790 V1.4.1 (09/05). Interaction channel for satellite distribution systems.

DVB BlueBook A054r4 (07/08). Interaction channel for satellite distribution systems (draft EN 301 790 V1.5.1 – DVB-RCS + M).

TR 101 790 V1.3.1 (09/06). Guidelines for the implementation and usage of the DVB interaction channel for satellite distribution systems.

DVB BlueBook A063r3 (10/08). Guidelines for the implementation and usage of the DVB interaction channel for satellite distribution systems (dTR 101 790 V.1.4.1).

DVB BlueBook A130 (11/08). Interaction channel for satellite distribution systems; Guidelines for the use of EN 301 790 in mobile scenarios (dTS 102 768 V.1.1.1).

EN 301 958 V1.1.1 (03/02). Digital Video Broadcasting (DVB); Specification of interaction channel for digital terrestrial TV including multiple access OFDM.

DVB BlueBook A073r1 (07/04). Interaction channel through General Packet Radio System (GPRS).

13.6 Middleware

For a full list of the MHP and GEM specifications for interactive TV middleware please visit www.mhp.org.

TS 102 523 V1.1.1 (09/06). DVB Portable Content Format (PCF) Specification 1.0.

TS 102 523 V1.1.1 schema DVB Portable Content Format (PCF) Specification 1.0 XML schema.

13.7 Content Protection and Copy Management

DVB BlueBook A094r2 (07/07). DVB content protection and copy management – A094r2 contains the complete set of core normative elements of the CPCM specification.

DVB BlueBook A129 (10/08). DVB-CPCM compliance framework.

13.8 Interfacing

ETS 300 813 V1 (12/97). DVB Interfaces to Plesiochronous Digital Hierarchy (PDH) networks.

ETS 300 814 V1 (03/98). Interfaces to Synchronous Digital Hierarchy (SDH) networks.

TR 100 815 V1.1.1 (02/99). Guidelines for the handling of ATM signals in DVB systems.

TS 101 224 V1.1.1 (07/98). Home Access Network (HAN) with an active Network Termination (NT).

TS 101 225 V1.1.1 (01/01). In-Home Digital Network (IHDN) Home Local Network (HLN).

EN 50221 V1 (02/97). Common interface specification for conditional access and other digital video broadcasting decoder applications.

R 206 001 V1 (03/97). Guidelines for implementation and use of the common interface for DVB decoder applications.

TS 101 699 V1.1.1 (11/99). Extensions to the common interface specification.

EN 50083-9 (2002). Interfaces for CATV/SMATV head ends and similar professional equipment.

TR 101 891 V1.1.1 (01/01). Digital Video Broadcasting (DVB); Professional interfaces: Guidelines for the implementation and usage of the DVB Asynchronous Serial Interface (ASI).

TS 102 201 V1.2.1 (01/05). Interfaces for DVB-IRDs.

13.9 Internet Protocol

TR 102 033 V1.1.1 (04/02). Architectural framework for the delivery of DVB-services over IP-based networks.

TS 102 034 V1.3.1 (10/07). DVB-IPTV 1.3: Transport of MPEG-2 TS-based DVB services over IP-based networks (and associated XML).

DVB Bluebook A086r7 (09/08). DVB-IPTV 1.4: Transport of MPEG 2 TS-based DVB services over IP-based networks (and associated XML) (dTS 102 034 V1.4.1).

TS 102 542 V1.2.1 (04/08). Guidelines for the implementation of DVB-IP Phase 1 specifications (c.f. TS 102 034).

TS 102 539 V1.2.1 (04/08). Carriage of Broadband Content Guide (BCG) information over Internet Protocol (IP).

TS 102 826 V1.1.1 (07/08). DVB-IPTV Profiles for TS 102 034.

DVB Bluebook A109 (02/07). DVB-HN (Home Network) Reference Model Phase 1.

DVB Bluebook A115 (05/07). DVB application layer FEC evaluations.

TR 102 824 V1.1.1 (07/08). Remote management and firmware update system for DVB IP services.

DVB Bluebook A128 (02/07). DVB-IP Phase 1.3 in the context of ETSI TISPAN NGN.

DVB Bluebook A132 (09/08). High-level technical requirements for QoS for DVB services in the home network.

TS 102 468 V1.1.1 (11/05). IP Datacast over DVB-H: Phase 1 specifications.

TR 102 469 V1.1.1 (05/06). IP Datacast over DVB-H: Architecture.

TS 102 470 V1.1.1 (04/06). IP Datacast over DVB-H: PSI/SI.

A079r2-1 (09/08). IP Datacast over DVB-H: PSI/SI (dTS 102 470-1 V1.2.1).

A079r2-2 (09/08). IP Datacast over DVB-SH: PSI/SI (dTS 102 470-2 V1.2.1).

TS 102 471 V1.2.1 (11/06). IP Datacast over DVB-H: Electronic Service Guide (ESG).

DVB Bluebook A099 Rev.1 (08/08). IP Datacast over DVB-H: Electronic Service Guide (ESG) (dTS 102 471 V1.3.1).

TS 102 592 V1.1.1 (10/07). IP Datacast over DVB-H: Electronic Service Guide (ESG) Implementation Guidelines.

TR 102 824 V1.1.1 (07/08). Remote management and firmware update system for DVB IP services.

TS 102 472 V1.2.1 (12/06). IP Datacast over DVB-H: Content Delivery Protocols (CDP).

TS 102 591 V1.1.1 (10/07). IP Datacast over DVB-H: Content Delivery Protocols (CDP) implementation guidelines.

TR 102 473 V1.1.1 (05/06). IP Datacast over DVB-H: Use Cases and Services.

TS 102 474 V1.1.1 (11/07). IP Datacast over DVB-H: Service Purchase and Protection.

TS 102 611 V1.1.1 (10/07). IP Datacast over DVB-H: Implementation guidelines for mobility.

A117r2-1 (09/08). IP Datacast over DVB-H: Implementation guidelines for mobility (dTS 102 611-1 V1.2.1).

A117r2-2 (09/08). IP Datacast over DVB-SH: Implementation guidelines for mobility (dTS 102 611-2 V1.2.1).

13.10   Conditional Access

ETR 289 V1 (10/96). Support for use of scrambling and Conditional Access (CA) within digital broadcasting systems.

DVB BlueBook A011r1 (06/96). DVB common scrambling algorithm: Distribution agreements.

TS 101 197 V1.2.1 (02/02). DVB SimulCrypt; Part 1: Head-end architecture and synchronization.

TS 103 197 V1.4.1 (09/04). Head-end implementation of SimulCrypt.

DVB BlueBook A045r4 (03/07). Head-end implementation of SimulCrypt (draft TS 103 197 V1.5.1).

TR 102 035 V1.1.1 (02/04). Implementation guidelines of the DVB simulcrypt standard.

13.11   Measurements

TR 101 290 V1.2.1 (01/01). Measurement guidelines for DVB systems.

TR 101 291 V1.1.1 (06/98). Usage of DVB test and measurement signaling channel (PID 0x001D) embedded in an MPEG-2 Transport Stream (TS).

TS 102 032 V1.1.1 SNMP MIB for test and measurement applications in DVB systems.

v2-tc.my; v2-tc.my MIB files containing the definition of "external" object referred in the TR 101 290 MIB.

v2-smi.my; v2-smi.my MIB files containing the definition of "external" object referred in the TR 101 290 MIB.

v2-conf.my; v2-conf.my MIB files containing the definition of "external" object referred in the TR 101 290 MIB.

7.SYSTEM-MIB.my text of chapter 7 of the draft TR 102 032 V1.1.1.

8.SIGNAL CHARACTERISTICS-MIB.my text of chapter 8 of the draft TR 102 032 V1.1.1.

9.TR101290-MIB.my text of chapters 9 of the draft TR 102 032 V1.1.1.

References

[Apa06a]   Maite Aparicio (editor). Wing TV. Services to Wireless, Integrated, Nomadic, GPRSUMTS & TV handheld terminals. D6: Common field trials report. November 2006. 86 p.

[Apa06b]   Maite Aparicio (editor). Wing TV. Services to Wireless, Integrated, Nomadic, GPRS-UMTS & TV handheld terminals. Wing TV Country field report. November 2006. 258 p.

[Apt06]   APT Recommendation on Guidelines for the Frequency Coordination for the Terrestrial Services at the Border Areas between Administrations. No. APT/AWF/2. Edition: Asia-Pacific Telecommunity. The APT Wireless Forum February 2006. Approved by the 30th Session of the APT Management Committee 18–21 September, 2006, Maldives.

[Bal07]   Gian Paolo Balboni. How Advertising Revenues Can Increase Profitability of DVB-H Services? Presentation, Deploying and Managing DVB-H, Gruppo Telecom Italia. London, September 27, 2007. 33 p.

[Bee07]   Karina Beeke. Spectrum Planning – Analysis of Methods for the Summation of Log–Normal Distributions. EBU Technical Review – October 2007. 9 p.

[Bmc07]   Mobile Broadcast Technologies. Link Budgets. BMCOFORUM, January 2007. 40 p.

[Bmc07b]   BMCOFORUM Spectrum Position for Mobile TV, March 2007. 4 p.

[Bou06]   Thibault Bouttevin (editor). D8 – Wing TV Measurement Guidelines and Criteria. Wing TV. Services to Wireless, Integrated, Nomadic, GPRSUMTS & TV handheld terminals. June 2006. 45 p.

[Bro02]   P. G. Brown, K. Tsioumparakis, M. Jordan, A. Chong. UK Planning Model for Digital Terrestrial Television Coverage. Research & Development. White Paper WHP 048. September 2002. 12 p.

[Dig05]   Television on a handheld receiver – broadcasting with DVB-H. Digital Terrestrial Television Action Group. Version 1.2, Grand-Saconnex, Geneva, Switzerland, 2005. 24 p.

[Dvb06]   Digital Video Broadcasting (DVB); DVB-H Implementation Guidelines. Draft TR 102 377 V1.2.2 (2006-03). 108 p.

[Dvb06b]   Digital Video Broadcasting (DVB); IP Datacast over DVB-H: Architecture. ETSI TR 102 469 V1.1.1 (2006-05). 28 p.

[Dvb07]   IP datacast over DVB-H: Implementation guidelines for mobility, DVB document A117, DVB organization, July 2007, pp. 18–19.

[Dvb08]   Broadcasting to Handhelds. DVB Fact Sheet, August 2008. Produced by the DVB Project Office. 2 p.

[Ecc04]   ECC Report 49. Technical Criteria of Digital Video Broadcasting Terrestrial (DVB-T) and Terrestrial – Digital Audio Broadcasting (T-DAB) Allotment Planning. *Electronic Communications Committee (ECC) Within the European Conference of Postal and Telecommunications Administrations (CEPT)*, Copenhagen, April 2004. 36 p.

[Erc99]   An Empirically Based Path Loss Model for Wireless Channels in Suburban Environments; Vinko Erceg, Larry J. Greenstein, Sony Y. Tjandra, Seth R. Parkoff, Ajay Gupta, Boris Kulic, Arthur A. Julius, Renee Bianchi; IEEE Journal on Selected Areas in Communications, Vol. 17, No. 7, July 1999.

[ETS04]   ETSI EN 301 192, V1.4.1 (2004-11). Digital Video Broadcasting (DVB); DVB specification for data broadcasting. European Telecommunications Standards Institute 2004. 78 p.

[ETS04a]   ETSI, "Digital Video Broadcasting (DVB); Framing Structure, Channel Coding and Modulation for Digital Terrestrial Television". ETSI standard, ETSI EN 300 744, November 2004.

[ETS04c]   ETSI, "Digital Video Broadcasting (DVB); Transmission System for Handheld Terminals". ETSI standard, ETSI EN 302 304, June 2004.

[ETS04d]   ETSI, "Digital Video Broadcasting (DVB); Transmission System for Handheld Terminals (DVB-H)", EN 302 304 v1.1.1, November 2004.

[ETS06]   ETSI, "IP Datacast over DVB-H; Set of Specifications for Phase 1", TS 102 468, 2006.

[ETS06b]   ETSI, "IP Datacast over DVB-H; PSI/SI", TS 102 470, 2006.

[ETS06c]   ETSI, "Digital Video Broadcasting (DVB); Specification for Data Broadcasting". ETSI standard, ETSI EN 301 192 v1.4.1, November 2004.

[Far06]   Gerard Faria, Jukka A. Henriksson, Erik Stare, Pekka Talmola. DVB-H: Digital Broadcast Services to Handheld Devices. *Proceedings of the IEEE*, Vol. 94, No. 1, January 2006. 16 p.

[Far07]   Gerard Faria. From Digital TV to Mobile TV. Presentation, CSTB 2007. 5 February 2007. 48 p.

[Gar07]   G. Gardikis, H. Kokkinis, G. Kormentzas. Evaluation of the DVB-H Data Link Layer. European Wireless 2007. 1–4 April 2007, Paris, France. 6 p.

[Goe02]   Roland Götz. Supporting Network Planning Tools II. Presentation, LS Telcom AG, 2002. 38 p.

[Gre06]   Emmanuel Grenier. DVB-H Radio Planning Aspects in ICS Telecom. White paper, ATDI. July 2006. 39 p.

[Had07]   Kamel Haddad. DVB-H in Denmark. Technical and Economic Aspects. Master's Thesis. Technical University of Denmark (DTU), Center for Information & Communication Technologies (CICT). September 10, 2007. 98 p.

[Hat80]   Masaharu Hata. Empirical Formula for Propagation Loss in Land Mobile Radio Services. IEEE Transactions on Vehicular Technology, Vol. VT-29, No. 3, August 1980. 9 p.

[Him06]   Heidi Himmanen, Ali Hazmi, Jarkko Paavola. Comparison of DVB-H link layer FEC decoding strategies in a mobile fading channel. The 17th Annual IEEE International Symposium on Personal, Indoor and Mobile Radio Communications (PIMRC'06). 1-4244-0330-8/06. IEEE 2006. 5 p.

[Him09]   Heidi Himmanen. On Transmission System Design for Wireless Broadcasting. Academic Dissertation. Department of Information Technology, University of Turku, Turku, Finland, 2009. 165 p.

[Imp07]   IP Datacast over DVB-H: Implementation Guidelines for Mobility. DVB Document A117 (BlueBook), July 2007. 26 p.

[ISO00]   ISO/IEC 13818-1. Information technology – Generic coding of moving pictures and associated audio information: Systems. 2000. 174 p.

[ITU07]   Recommendation ITU-R P.1546-3. Method for point-to-area predictions for terrestrial services in the frequency range 30 MHz to 3000 MHz. 2007. 57 p.

[Jeo01]   Minseok Jeong, Bomson Lee. Comparison between Path-Loss Prediction Models for Wireless Telecommunication System Design. 0-7803-7070-8/01. IEEE 2001. pp. 186–189.

[Jok05]   Heidi Joki, Jarkko Paavola, Valery Ipatov. Analysis of Reed–Solomon Coding Combined with Cyclic Redundancy Check in DVB-H Link Layer. ISWCS05. 0-7803-9206-X/05. IEEE 2005. pp 313–317.

[Jos07]   Wout Joseph, Emmeric Tanghe, Daan Pareit, Luc Martens. Building Penetration Measurements for Indoor Coverage Prediction of DVB-H Systems. 1-4244-0878-4/07. IEEE 2007. pp. 3005–3008.

[Kru05]   Stefan Krueger. DVB-H Pilot Berlin. Presentation. T-Systems International GmbH Media&Broadcast, Broadcast Network & Services. September 2005. 23 p.

[Lee86]   William C.Y. Lee. Elements of Cellular Mobile Radio System. IEEE Transactions on Vehicular Technology, Vol. VT-35, No. 2, May 1986. pp. 48–56.

[Lim99]   Limits of Human Exposure to Radiofrequency Electromagnetic Fields in the Frequency Range from 3 kHz to 300 GHz. Minister of Public Works and Government Services, Canada 1999. Cat. H46-2/99-237E. ISBN 0 662 28032 6. 40 p.

[Lun06]   Janne Lundberg. A Wireless Multicast Delivery Architecture for Mobile Terminals. Dissertation for the Degree of Doctor of Science in Technology. Helsinki University of Technology (Espoo, Finland) on the 15th of May, 2006. 138 p.

[Mäk05]   Juri Mäki. Finnish Mobile TV Pilot Results. August 30th, 2005. Research International Finland. 13 p.

[Mar05]   Peter Marshall. Digital Television Project. Advanced Receiver Techniques with Emphasis on Portable TV Reception. Issue 1.1. DTG Management Services Ltd, March 2005. 65 p.

[Maz07]   Joan Mazenc. Multi protocol Encapsulation – Forward Error Correction (MPE-FEC). ETUD/Insa. Toulouse, France, January 9, 2007. 7 p.

[Mil06]   Davide Milanesio (editor). Wing TV. Services to Wireless, Integrated, Nomadic, GPRSUMTS & TV Handheld Terminals. D11 – Wing TV Network Issues. May 2006. 140 p.

[Min99]   Limits of Human Exposure to Radiofrequency Electromagnetic Fields in the Frequency Range from 3 kHz to 399 GHz. Safety Code 6. Environmental Health Directorate, Health Protection Branch. Publication 99-EHD-237. Minister of Public Works and Government Services, Canada 1999. ISBN 0-662-28032-6. 40 p.

[Myr06]   Myron D. Fanton. Analysis of Antenna Beam-tilt and Broadcast Coverage. ERI Technical Series, Vol. 6, April 2006. 3 p.

[Nok05]   Technical data sheet of Nokia N-92. Nokia, 2005. 1 p.

[OMA08]   OMA: "Mobile Broadcast Services", Draft Version 1.0–03 November 2008, Open Mobile Alliance, OMA-TS-BCAST_Services-V1_0-200801103-D.

[OMA08b] OMA: "Service Guide for Mobile Broadcast Services", Draft Version 1.0–28 October 2008, Open Mobile Alliance, OMA-TS-BCAST_Service_Guide-V1_0-20081028-D.

[Paa07]   J. Paavola, H. Himmanen, T. Jokela, J. Poikonen, V. Ipatov. The Performance Analysis of MPE-FEC Decoding Methods at the DVB-H Link Layer for Efficient IP Packet Retrieval. IEEE Transactions on Broadcasting, Vol. 53, No. 1, March 2007. pp. 263–275.

[Pek05]   Stuart Pekowsky, Khaled Maalej. DVB-H architecture for mobile communications systems. RF Design, April 2005. pp 36–42.

[Pen08]   Jyrki T.J. Penttinen. Field Measurement and Analysis Method for DVB-H Terminals. *The Third International Conference on Digital Telecommunication. ICDT 2008*, 29.6–5.7.2008, International Academy, Research, and Industry Association (IARIA), Bucharest, Romania. 6 p.

[Pen08b] Jyrki T.J. Penttinen. CAPEX and OPEX Optimisation in Function of DVB-H Transmitter Power. *The Third International Conference on Digital Telecommunication. ICDT 2008*, 29.6.–5.7.2008, International Academy, Research, and Industry Association (IARIA), Bucharest, Romania. 6 p.

[Pen08c] Jyrki T.J. Penttinen. The Simulation of the Interference Levels in Extended DVB-H SFN Areas. *The Fourth International Conference on Wireless and Mobile Communications.* International Academy, Research, and Industry Association (IARIA), 2008. pp. 223–228.

[Pen08d] Jyrki T.J. Penttinen. The SFN gain in non-interfered and interfered DVB-H networks. *The Fourth International Conference on Wireless and Mobile Communications.* International Academy, Research, and Industry Association (IARIA), 2008. pp. 294–299.

[Pen08e] Jyrki T.J. Penttinen. DVB-H Performance Simulations in Dense Urban Area. *The Third International Conference on Digital Society.* International Academy, Research, and Industry Association (IARIA), 2009. 6 p.

[Pen09]   Jyrki T.J. Penttinen. "DVB-H Performance Simulations in Dense Urban Area". *The Third International Conference on Digital Society (ICDS 2009).* February 1–7, 2009, Cancun, Mexico. International Academy, Research, and Industry Association (IARIA). 6 p.

[Pen09b] Jyrki T.J Penttinen, Eric Kroon. "MPE-FEC Performance in Function of the Terminal Speed in Typical DVB-H Radio Channels". *IEEE International Symposium on Broadband Multimedia Systems and Broadcasting (BMSB).* 13–15 May, 2009, Bilbao, Spain. 6 p.

[Pir99]   Riku Pirhonen, Tapio Rautava, Jyrki Penttinen. TDMA Convergence for Packet Data Services. IEEE Personal Communications. June 1999. pp. 68–73.

[Ple08]   D. Plets, W. Joseph, L. Verloock, E. Tanghe, L. Martens, E. Deventer, H. Gauderis. Influence of Reception Condition, MPE-FEC Rate and Modulation Scheme on Performance of DVB-H. IEEE Transactions on Broadcasting, Vol. 54, No. 3, September 2008. pp. 590–598.

[Sat06]   Claus Sattler. Mobile Broadcast Business Models: A State-of-the-Art Study. BMCOFORUM, November 2006. 36 p.

[Sat07]   Claus Sattler. Mobile Broadcasting: Trends and Challenges. LS Summit. 20. June 2007, Lichtenau. BMCOFORUM. 23 p. !!! see above links for link budget etc bmcoforum docs!!!

[Ung07]   Peter Unger, Thomas Kuerner. Optimizing the Local Service Areas in Single Frequency Networks. COST 2100 TD(08)467. Wroclaw, Poland, 2008, February 6–8, 7 p.

[Var04]   Jani Väre and M. Puputti, "Soft Handover in Terrestrial Broadcast Networks", in *Proc. of the 2004 IEEE International Conference on Mobile Data Management*, Berkeley, CA, USA, January 2004, pp. 236–242.

[Var06]   Jani Väre, J. Alamaunu, H. Pekonen and T. Auranen, "Optimization of PSI/SI Transmission in IPDC over DVB-H Networks", in *Proc. of the 56th Annual IEEE Broadcast Symposium*, Washington, DC, USA, September 2006.

[Zha06]   Yue Zhang, John Cosmas, Maurice Bard, Yong-Hua Song. Diversity Gain for DVB-H by Using Transmitter/Receiver Cyclic Delay Diversity. IEEE Transactions on Broadcasting, Vol. 52, No. 4, December 2006. pp 464–474.

Internet Links

DVB organization – www.dvb.org.

DVB, "IP Datacast over DVB-H: Set of Specifications for Phase 1- "http://www.dvb-h.org/PDF/a096. tm3409r1.cbms1471r3.IPDC_Phase_1_Specs.pdf"\t "_blank" A096 ", 2005.

DVB-H blue book/standards - http://www.dvb.org/technology/standards/index.xml.

DVB-H organization – www.dvb-h.org.

Encryption - www.cm-la.com.

ETSI – www.etsi.org.

ETSI, "DVB-H Validation Task Force Report", "http://www.dvb-h.org/PDF/Validation%20Task%20Force% 20Report%20-%20Tr102401.V1.1.1.pdf"\t "_blank" TR 102 401 v1.1.1, May 2005.

ETSI, "IP Datacast over DVB-H: PSI/SI", "http://www.dvb-h.org/PDF/ts_102470v010101p.pdf"\t "_blank" TS 102 470 v1.1.1, April 2006.

ITU – www.itu.org.

Nokia's Mobile TV portal – www.nokia.com/mobiletv.

Nokia's MobileTV, http://www.mobiletv.nokia.com.

Open Mobile Alliance (OMA), www.openmobiealliance.org.

T. Jokela and Jani Väre, "Simulations of PSI/SI Transmission in DVB-H Systems", "http://www.ieee.org/ organizations/society/bt/BMS07/07bmsindex.html" 2007 IEEE International Symposium on Broadband Multimedia Systems and Broadcasting, Orlando, FL, USA, March 2007.

Index